# DRYLAND RESOURCES AND TECHNOLOGY

Volume 7, 1992

*Editor-in-Chief*

**L. L. Somani**

Rajasthan College of Agriculture, Udaipur.

*Co-Editor*

**R. L. Shyampura**

Regional Centre NBSS & LUP

UDAIPUR – 313 001

&

**M. C. Mundra**

CCS Haryana Agricultural University

HISAR – 125 004

SCIENTIFIC PUBLISHERS

Published by:

**SCIENTIFIC PUBLISHERS (INDIA)**
5A, New Pali Road, P.O. Box 91
Jodhpur 342 001, India
E-mail: info@scientificpub.com
Website: www.scientificpub.com

**Print : 2021**



ISBN : 978-93-88449-51-9
eISBN: 978-93-88449-52-6

Printed in India

# CONTENTS

# *Contents of Previous Volumes*

**VOLUME 1**

Fauna of The Great Indian desert. *M.L. Roonwal*

Water reclamation for potable use. *Oliver O. Hart*

Desalination of water. *Alam Singh and U.S. Sharma*

Unsaturated flow in an arid environment. *Clinton M. Case*

Solar and wind energies utilization. *H.P. Garg & N.D. Kaushika*

Reclamation and use of ravine lands. *R.K. Gupta & M.C. Prajapati*

Pasture development. *K.C. Kanodia & B.D. Patil*

Economic and medicinal plants of Indian desert. *Vijendra Singh and R.P. Pandey*

**VOLUME 2**

Field heterogeneity and its hydrological implications. *M.L. Sharma (Australia)*

Phytotoxicity assessment of some industrial and combined domestic wastes. *Samia G. Saad & Ahmed A. Hamza (Egypt)*

Management of dryland landscapes. *John Zwar (Australia)*

Greenhouse engineering for desert cultivation. Pitam Chandra, *C.P. Bohra & R.C. Maheshwari (India)*

Utilization of marginal water in arid zones. *Gideon Oron (Israel)*

Estimation of runoff volume for agriculture in arid lands. *J. Ben-Asher, G. Oron & B.J. Button (Isreal)*

Isotope hydrology. *H. Chandrasekharan (India)*

Model studies for design of shelter belt. *V.P. Gupta (India)*

Resources for planning in the arid areas of Rajasthan : A case study of Malani region. *T.S. Chouhan (India)*

Forages for arid lands. *R.S. Paroda & D.S. Jatasra (India)*

Regional Cooperating Editors.

**VOLUME 3**

New Source of Irrigation Water for Arid Agriculture in Kuwait. *G.A. Al-Nakshabandi & B. Salem*

Stress Physiology and plant cell cultures. *J.J. Oertli & A. Mozafar*

r C and r A (Mathematical synthesis of the ionic characteristics of a water solute). *Crescencio – Ignacio Villalobos*

Management of mineral and power resources in dryland of Rajasthan. *T.S. Chouhan*

Dryland agriculture and land development practices in red soils of semi-arid regions. *T.C. Channappa*

Human resources for economic development planning in drylands of Rajasthan. *T.S. Chouhan*

Irrigated forage : Production for the world's largest integrated dairy farm. *Richard Cates Jr. & J.S. Zeinstra*

The Malqaf : A traditional cooling and ventilation system. *Hassan Fathy & Abd-el-rahman Ahmed Sultan*

A non-tracking high-temperature solar collector for refrigeration Applications. *K.L.J. Webb, T.S. Prasad & A.G. Williamson*

**VOLUME 4**

Plant resources of arid lands in India. *R.S. Paroda & S. Mauria*

Alternate farming systems for drylands of semi-arid tropics of India. *R.P. Singh & S.P. Singh*

Water by electrodialysis through photovoltaic power for desert *survival. K.P. Govindan*

Remote sensing of crop water stress. *Usha K. Chopra & Naween Kalra*

Dynamics of bare soil evaporation and its modifications. *R.S. Malik & R.K. Malik*

Soil, water and crop management practices for improving productivity of drylands of semi-arid tropics of India. *R.P. Singh*

Agromelioration of sodic soils. *L.L. Somani*

Tolerance of oilseed to salinity and alkalinity - a review. *D. Kumar*

Livestock development in drylands of Rajasthan. *T.S. Chouhan*

The application of new sprinkler systems- soil-mechanical and agricultural aspects. *Peter Wolff*

Subsurface irrigation-a modern irrigation and fertilization technique for increasing yield. *Hans Jankowiak*

Chances and risks of biotechnology for agricultural production in the tropics and subtropics. *Josef Schell & Jorge Mayer*

## VOLUME 5

Operational research in dryland agriculture. *T. Vishnu Murthy*

Dryland afforestation. *G.N. Gupta*

Application of remote sensing techniques in dryland Agriculture. *Baldev Sahai & Ajai*

Reclamation of alkali soils through green manuring. *L.L. Somani, et al.*

Problems and prospects to industrial development in the desert region of India. *T.S. Chouhan & K.C. Gujar*

Research efforts to mitigate the adverse effects of drought. *R.P. Singh*

Alternate cropping system for arid or semi-arid regions. *M.S. Shaktawat & H.G. Singh*

Soil fertility management and fertiliser-use in dryland agriculture. *N.K. Kalyanasundaram*

Soil and moisture conservation for drylands. *G.L. Maliwal*

Environmental impact of flood in dry zone of India. *K.N. Joshi*

## VOLUME 6

Diseases of arid-zone seed-spices: Present status and future strategies for their management. *J.P. Agnihotri*

Crops and cropping system for drylands. *S.S. Rathore & I.N. Gupta*

Effectiveness of soil management in relation to rainfall on crop productivity in sandy soils of India. *K.P.R. Vittal*

Tillage in coarse textured soils. *M.R. Chaudhary*

Potential grasses for arid zone. *M.S. Yadav*

Compaction of sandy soils and plant growth. *I.J. Gulati et al.*

Pitcher irrigation technique for arid and semi-arid zones. *Sarwan K. Dubey*

Oilseeds production in rainfed areas-constraints and opportunities. *M.P. Sahu*

## VOLUME 8

Effect of water deficit on symbiotic nitrogen fixation in legumes. K. Swaraj & N.R. Bishnoi

Characterization of resources and their efficient management for sustainable productivity of rainfed crops grown on black clay soils. R.A. Sharma

Crop production with sodic water. H.D. Yadav & V. Kumar

Present status of plant virological researches for the arid and semi-arid tracts of Rajasthan and future strategies. J.P. Agnihotri

Amelioration of alkali soils through Fly Ash. *L.L. Somani et al.*

An integrated approach towards increased and sustained production from drylands of eastern uttar Pradesh. *S.R. Singh, R.P. Singh & Y. Singh*

Prospects of increasing fertiliser use efficiency in dryland farming. *M.P. Sahu*

An assessment of genetic variability in *Setaria italica* L. *R.P. Gupta, U.C. Sharma & S.R. Maloo*

An integrated approach towards increased and sustained production from drylands of Easters Utter Pradesh. *R.P. Singh et al.*

Management options for vertisols and associated soils of central and peninsular India. *R.A.Sharma*

Dryland Resources & Technology, 7 (1992)
Editors : L.L. SOMANI, R.L. SHYAMPURA
& M.C. MUNDRA


# Energetics of Country Plow/tractor use in Millets-castor Bean Rotation on Medium Deep Alfisols under Dryfarming

*C. Sriram and T. Vishnu Murthy*[*]
[*]CRIDA, Saidabad,
Hyderabad–500 659

## ABSTRACT

Under dryland agriculture, tillage-both primary and secondary- helps in increasing crop production due to improvement in soil profile characteristics. The energy utilisation for these tillage operations was not monitored over time for the improved crop technology. During 1982-86, a gradient of tillage energy was superimposed using tractor and country plow on improved cultural practices under six treatments in a four replicated trial in millets-castor bean rotation. The output/input ratio in castor bean was lower than that of millets. Both input energy and rainfall significantly contributed towards output energy. The total animal, machinery and human energy was almost quantitatively similar under tractor and country plow. With tractor tillage, diesel contributed about 20% of input energy. Among primary tillage, sowing and interculture, the former two contributed positively towards output while the latter contributed negatively. The tillage at sowing was of primary importance. The requirement of energy was reduced with time for primary tillage, sowing and summer tillage under both country plow and tractor. Due to high spatial and temporal variation of rainfall and soil types in drylands, mechanisation is required for speedy completion of farm operations.

## INTRODUCTION

Under dryland agriculture, tillage has signifidcant role due to the rainfed nature of crop productivity. The positive effects of tillage on infiltration, increased residual moisture and effective smothering of weeds are well known. Information on energetics of the systems are, however, lacking.

In the improved technology for higher crop productivity the seed, fertiliser, plant protection, etc. are standardised to a large extent, for various agro-climatic regions. Energy aspects of method of tillage by country plow or tractor, received less attention. At present, the energy involvement in the above methods of tillage were quantified and energy requirements over time were evaluated. This study, adds to the understanding of the energy management under dry farming conditions.

## MATERIALS AND METHODS

In the medium deep alfisols of Gunegal Research Farm of Central Research Institute for Dryland Agriculture, Hyderabad, an experiment with varying degrees of energy input in tillage, super-imposed on recommended agronomic practices, was carried out during 1982- 86. The experimental lay out was vividly described earlier (Sriram and Vishnumurthy, 1987). The treatmental details are presented in results hereunder (Table-1). Equivalents for direct and indirect sources of energy were calculated as per standard methods (Mittal et al. 1986).

At present, the energy relations were quantified from these gradients from relevant contrasts in the treatments in the millets-castor bean rotation. The input-output energies were related statistically by regression analysis..

## RESULTS

Under dryland agriculture, the research work carried out over years, had identified the fertiliser dose, crop and its cultivar, agronomic practices like sowing time, interculturing, etc. For medium deep alfisols of Telengana region, a millet-castor bean rotation was efficient. In this rotation, a gradient of input energies were introduced with tractor/country plow in an experiment described earlier(Sriram and Vishumurthy, 1987). In that research paper, the effects of treatments on grain and total biomass were described. At present, these energy relations were described in both millets and castor bean.

Table 1 : Input energy and output/input energy ratios of total biomass of millets and castor bean

| S No. | Treatment | Input energy (MJ/ha) | Output/ input energy ratio of total biomass |
|---|---|---|---|
| | **Millets** | | |
| 1) | One MB ploughing (TD) + one blade harrowing (TD) + sowing by MF seed drill (TD) | 6115 | 7.94 |
| 2) | Twice blade harrowing (TD) + sowing by MF seed drill (TD) | 5865 | 8.90 |
| 3) | Twice six tyne cultivator (TD) + sowing by MF seed drill (TD) | 5839 | 10.23 |
| 4) | Country plow + sowing by hand + two blade harrowings (interculture) and two hand weedings | 5044 | 10.95 |
| 5) | Chemical weed control + sowing by MF seed drill (TD) and no hand weeding and interculture | 4627 | 7.12 |
| 6) | Chemical weed control + sowing by hand + one blade harrowing and hand weeding | 5342 | 7.76 |
| | SEm | 140.9 | 0.486 |
| | CV (%) | 5.7 | 12.3 |

Table contd............

| S. No. | Treatment | Input energy (MJ/ha) | Output/input energy ratio of total biomass |
|---|---|---|---|
| | **Castore bean** | | |
| 1) | One MB plowing (TD) + one blade harrowing (TD) + sowing by MF seed drill (TD) | 7593 | 5.54 |
| 2) | Twice blade harrowing (TD) + sowing by MF seed drill (TD) | 7221 | 6.48 |
| 3) | Twice six tyne cultivator (TD) + sowing by MF seed drill (TD) | 7268 | 7.14 |
| 4) | Country plow + sowing by hand + twice blade harrowing and hand weeding | 6424 | 6.51 |
| 5) | Direct sowing by MF seed drill (TD) and no hand weeding and blade harrowing | 5360 | 4.99 |
| 6) | Direct sowing by MF seed drill (TD) + two hand weedings without blade harrowing | 5974 | 5.99 |
| | SEm | 81.9 | 0.35 |
| | CV (%) | 2.8 | 12.6 |

MB Mould board; TD = Tractore drawn; MF = Massey-Ferguson

The results of the input energy and ratio of output/input energy are presented for millets and castor bean in Table-1. In general, castor bean, a long duration crop of 120-150 days, consumed more energy than the millets (short duration crops of 70-100 days). However, the energy recovery was higher in the cereal millets than the oil seed castor beans. As the mechanical/animal/human energy was reduced as in later treatments the total input energy was also reduced. These three forms of energy were the main sources of tillage in the dryland agriculture. In general, higher output of energy could be obtained with higher tillage.

A perusal at the energy use patterns (Table-2) in castor bean and millets, had showed almost similar energy use. There was marginal difference in the use of seed energy and machinery energy. Diesel use made a significant difference in the total energy with tractor. Human energy and animal energy requirements were considerably reduced with the use of tractor.

Table 2 : Patterns of total energy utilisation in millet and castor bean under country plow and tractor cultivation

| Form | Percent energy utilisation | | | |
|---|---|---|---|---|
| | Millets | | Castor bean | |
| | Country plow | Tractor | Country plow | Tractor |
| Human energy | 24 | 15 | 22 | 16 |
| Animal energy | 10 | 5 | 8 | 4 |
| Machinery energy | 12 | 11 | 12 | 11 |
| Diesel | 0 | 24 | 0 | 20 |
| Seed | 2 | 2 | 6 | 4 |
| Fertiliser | 50 | 41 | 50 | 43 |
| Pesticides | 2 | 2 | 2 | 2 |
| Total | 100 | 100 | 100 | 100 |

In drylands, both the input energy and rainfall are significant in the resultant of the output energy due to rainfed dependency for production. Following is the equation describing this relationship over the crops and seasons:

Output (MJ/ha) =

8.1** Input (MJ/ha) + 18.5** Rainfall (mm) - 18890

(0.67) (15.183)

Multiple R = 0.961

where, the figures in parantheses are SE of the coefficient and ** coefficients are significant at 0.01 probability. The input, rainfall and output had a range of 2159-8565 MJ/ha, 293-781 mm and 7146-61848 MJ/ha, and a mean of 5787 MJ/ha, 448 mm and 36236 MJ/ha, respectively.

Among different components, the energy source was almost same for fertiliser, pesticide and seed for a crop. The important variations occurred in the input are primary tillage, sowing and interculture. The variation in input energy through primary tillage, sowing and interculture were 370-1632, 123-581 and 363-1158, with a mean of 950, 339 and 834 MJ/ha, respectively. The relationship between output and input energy through the above three sources is described below:

Output (MJ/ha) = 35239 + 8.8 Primary tillage (MJ/ha)

(40.03)

+23.55 Sowing (MJ/ha)

(111.62)

–10.598 Interculture (MJ/ha)

(20.8390)

Multiple R = 0.627

where, figures in parantheses are SE.

A negative trend with interculture indicates the effect of reduction in output energy with increasing density of weed population. Increase in primary and sowing energy, increased the production.

Over a period of 5 years, the energy use was monitored in both millets (Fig. 1) and castor bean (Fig.2). In general, it was seen that with time there was a significant linear reduction in energy requirement, for both primary tillage and sowing. The coefficients are presented in Table-3. Both in millets and castor bean, sowing consumed about 1/3 of energy in comparison to primary tillage. For these two operations, the amounts of energy utilised in the crops was almost similar. The reduction in requirement of energy for sowing varied from 28.2 to 33.4 MJ/ha/year with an average of 30.5 MJ/ha/year. In primary tillage, the decrease ranged from 75.5-127.5 MJ/ha/year with an average of 94.3 MJ/ha/year. The reduction in primary tillage requirement was significantly superior in castor bean with tractor tillage. Among others, the differences were marginal.

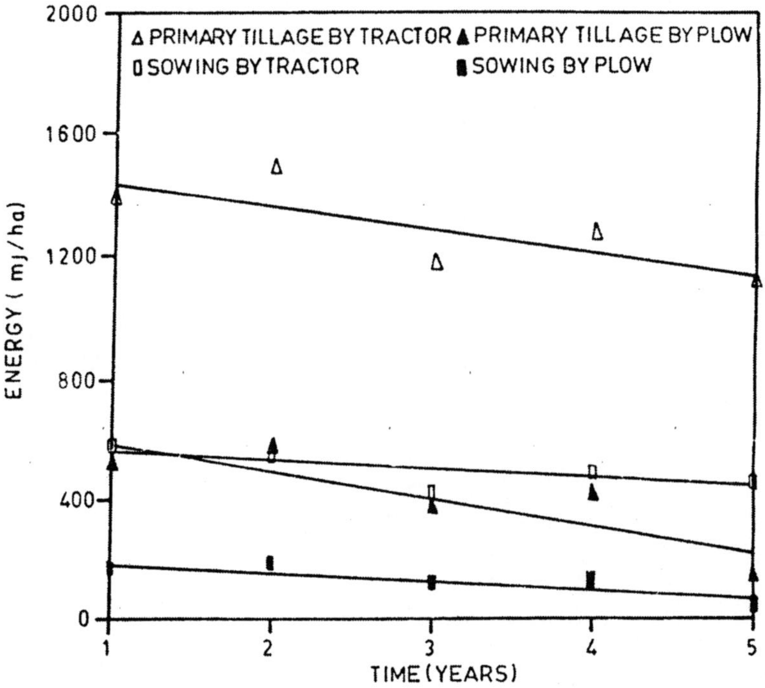


**Fig. 1.** Energy use over time in millets

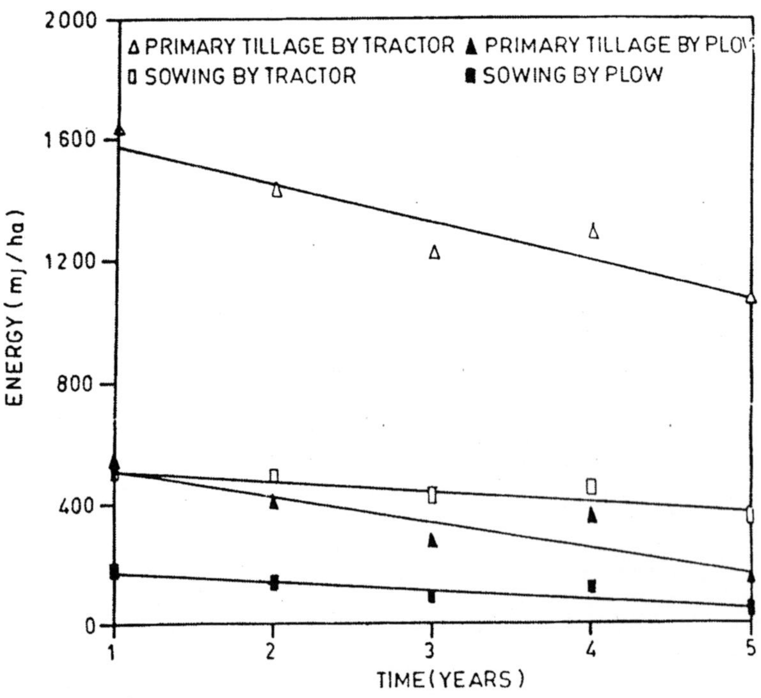


**Fig. 2.** Energy use over time in castor

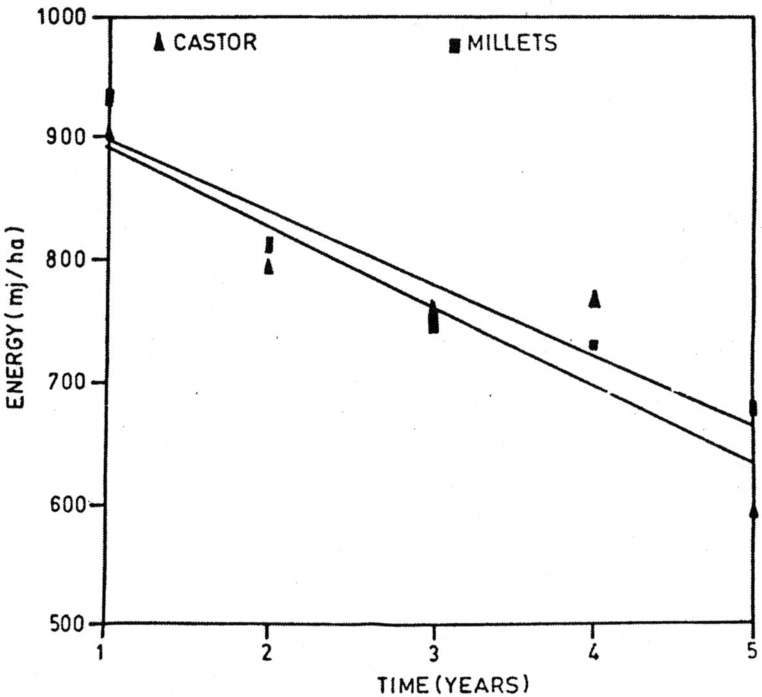


**Fig. 3.** Energy use for summer tillage with tractor

During off-season after the cessation of monsoon period, the soil becomes compact and hard due to receding soil moisture with soil bulk density reaching upto 1.8-1.9 g/cc. To eliminate the weed growth and open the soil for better infiltration, summer tillage is essential. This is more feasible with only higher draft power availability. The energy use for summer tillage over time is presented in Fig.3. Statistical analysis given in Table-3 had shown that for both the crops the differences in slope and intercept of the line were marginal. Thus, the summer tillage was not affected by the previous crops. The energy requirement for summer tillage was confounded by the soil effect. The mean was 61.3 MJ/ha/year.

Table 3 : Relationships between energy use and time (X, independent variable) for primary tillage and summer tillage, sowing using tractor and country plow in millets and castor bean

| Energy use by dependent variable (Y) | Slope | Intercept | Standard error | Correlation coefficient |
|---|---|---|---|---|
| **In millets using country plow** | | | | |
| Primary tillage | 673 | –90.1 | 30.12 | –0.865* |
| Sowing | 225 | –30.1 | 10.03 | –0.866* |
| **In millets using tractor** | | | | |
| Primary tillage | 1508 | –75.5 | 33.84 | –0.790 |
| Sowing | 592 | –30.3 | 15.27 | –0.753 |
| **In millets using tractor** | | | | |
| Summer tillage | 958 | –58.5 | 10.83 | –0.952** |
| **In castor bean using country plow** | | | | |
| Primary tillage | 589 | –84.2 | 22.75 | –0.906* |
| Sowing | 197 | –28.2 | 7.61 | –0.906* |
| **In castor bean using tractor** | | | | |
| Primary tillage | 1706 | –127.5 | 28.07 | –0.934** |
| Sowing | 546 | –33.4 | 10.23 | –0.883* |
| **In castor bean using tractor** | | | | |
| Summer tillage | 955 | –64.2 | 16.28 | –0.916* |

** * significant at 0.01 and 0.05% probability

## CONCLUSIONS

The energy use analysis under dryland agriculture in millet-castor bean crop rotation in medium deep alfisols indicated that the basic energy use difference under tractor/country plow use occurs basically due to diesel utilisation, a non-renewable energy source. With the use of tractor a saving in human and animal energy was recorded.

The requirement of energy, in general, for primary tillage, sowing and summer tillage decreased linearly with time, 94, 31 and 61 MJ/ha/year, respectively even under highly varying rainfall patterns. However, the amount of rainfall significantly contributed along with input energy towards the output energy. The contribution of rainfall was more than double that of input energy. Among different tillage energies, primary tillage and sowing contributed positively to total output while the interculture was related negatively. Wherever, higher energy was used for interculture, the output energy was lower. In general, it is seen that use of any tillage continuously on the land makes the soil more open, thus, reducing the energy requirement in later years.

In view of the compactness of the soil profile, a judicious mixed use of tractor and country plow are suggested. The summer and primary tilages need to be carried out with tractor when a country plow cannot break open the soil. The sowing can be carried out with improved seed drills. Multi-purpose implements which can be used for sowing/interculture/fertiliser application are beneficial. However, if a large area is to be seeded in short time, the need for tractor seed drills with higher field capacity is inevitable due to high spatial and temporal variation of rainfall in drylands.

The two main features arise out of this study are: the total of human, animal and machinery energy was more or less same with country plow and tractor; and there was significant reduction in tillage energy with time. These observations induces a significant possibility to replace traditional systems with improved mechanisation for efficient coverage of land under present day trends of decrease in the cost and increase in production of non-renewable energy like diesel. Thus, mechanisation could find a significant role under dryfarming for increasing and production stabilising.

## ACKNOWLEDGEMENTS

Facilities received from Director and staff, Central Research Institute for Dryland Agriculture, Hyderabad during experimentation are gratefully acknowledged.

## REFERENCES

Mittal, J.P., Sodhi, K.S. and Dhawan, K.C. (1986) Research Manual on Energy Requirements in Agricultural Sector. Published by Co-ordinating Cell, All India Co-ordinated Research Project on Energy Requirements in Agricultural Sector, Ludhiana, pp 136.

Sriram, C. and Vishnumurthy, T. (1987) Minimum tillage requirements for castor-sorghum/pearl millet rotation. J. Agric. Engg. ISAE 24 (3) 239-244.

Dryland Resources & Technology, 7 (1992)
Editors : L.L. SOMANI, R.L. SHYAMPURA
& M.C. MUNDRA


# Weed Management in Dryland Crops

*N.T. Yaduraju and K.N. Ahuja*
Division of Agronomy, IARI,
New Delhi–110 012

## ABSTRACT

Weeds are as important in dryland situations as they are in irrigated conditions. During the years of favourable weather condtions, competition by weeds may be more for nutrients and space. However, due to their better adaptive mechanisms, weeds invariably compete better than crops under adverse soil and weather condtions. Hence good management practices are a must to tilt the balance of weed-crop competition in favour of crops. Heavy losses in crop yields occur due to unchecked growth of weeds. Although not much information is available regarding yield losses in farmer's field, it is generally believed that weed control is not taken up by the farmers with earnest due to sheer ignorance, lack of availability of labour or inhospitable soil or climatic conditions.

Tillage with the objective to control weeds is widely practised in drylands. Where minimum cultivation is required to conserve the soil, weeds have to be taken care of either by manual or chemical methods. Crop husbandry practices like selection of competitive crop or crop cultivar, manipulations in plant population and planting geometry, intercropping, crop rotation etc. have good potential in weed suppression which need to be exploited. There is much meaning in the saying 'Good crop is the best weed killer'. Herbicides have their own advantages but their use is currently limited because of socio-economic and technological reasons. The method needs to be popularised through proper training of farmers and extension personnel. The high cost of herbicides can be reduced by lowering the herbicide dose. This can be achieved by combining herbicides with other methods of weed control or resorting to band, instead of broadcast, application over the crop row only. The weeds in the untreated space between the rows will have to be controlled by intercultivation operations.

Control of weeds through integration of mechanical, cultural and chemical method is by far the best method of weed control, as such weed management programme exploits the strengths and compensates for the weaknesses of individual methods. Special weed problem in dryland areas are discussed and control measures given.

## INTRODUCTION

Despite all efforts to control them, weeds continue to inflict heavy losses in crop production. Considerable amount of time, energy and money is spent in our fight against this pest problem. It is almost impossible to estimate losses in crop yield due to weeds without creating an artificial environment to separate their effects from those caused by insects, diseases, soil and atmospheric conditions. Nevertheless, it is generally accepted that, world wide some 10% loss of agricultural production can be attributed to the competitive effect of these plants. World wide in cereals alone, this represents a loss of more than 155.5 million tons. There is of course considerable variation between different countries, between different crops and even between the same crop in different situations. In India according to a very conservative estimate, the annual crop losses due to pests amount to a staggering Rs. 15,000 crores (Rangaswamy, 1983), of which, about one-third loss is due to weeds. Apart from the ill-effects of weeds reported in several text books, the weed problem in dryland agriculture assumes special significance because of following factors:

1. There is little time between land preparation and sowing of crops. Quite often the crop is planted without proper field preparation at all. Weeds particularly perennial weeds are bound to take an upperhand under such circumstances.

2. The crops are grown under limited supply of water and are often grown under stored soil moisture. By virtue of their deep and extensive root system weeds invariably take a upper hand over crops. It is made worse by the fact

that weeds in general have higher transpiration coefficient than crop plants.

3. The magnitude of competition by weeds is dependent on population and health and vigour of crop plants. Suboptimal input supply and management practices, which are common in dry farming system, would benefit weeds more than crop plants.

4. The socio-economic condition of the farmer is also of considerable importance. Many of these farmers actually ill afford the modern crop production technology.

5. The returns are poor and unpredictable. The benefit cost ratio is invariably low as compared to irrigated areas.

Like research in other fields, there has been a conspicuous bias towards irrigated agriculture as far as research in the field of weeds and their control is concerned. In this publication, however, an attempt has been made to review the subject with particular reference to dry land agriculture.

## WEED FLORA

Diverse weed flora is seen in dryland areas. Information through weed scientists was collected from different parts of the country (Yaduraju and Ahuja, unpublished) and the weed flora as reported by them is given below in alphabetical order.

### a) Kharif Weeds

*Andropogon aciculatus, Alhagi camelorum, Acanthosperum hispidum, Borreria hispida, Brachiaria* sp., *Bidens pilosa, Blumea lacera, Boerhavia diffusa, Commelina diffusa, Celosia argentea, Cyanotis axillaris, Croton sparsiclorus, Corchorus acutangulus, Cenchrus ciliaris, Cymbopogan* sp., *Crosophora* sp., *Dinebra retroflexa, Desmostachya bipinnata, Dactyloctenium* sp., *Eupatorium odoratum, Evolvulus nummularius, Eclipta* sp., *Eragrostis* sp. *Galinsoga parviflora, Gomphrena* sp., *Gynandropsis pentaphylla, Ludwigia octovalvis, Imperata cylindrica, Leucas aspera, Ludwigia parviflora, Lagascea mollis, Mimosa pudica, Malva* sp., *Panicum* sp., *Paspalum* sp., *Parthenium hysterophorus, Physalis minima, Pluchea lanceolata, Phyllanthus niruri, Portulaca* sp., *Rotala rotundifolia, Saccharum* sp., *Sida acuta, Trian-*

*thema* sp., *Tridax procumbens, Tribulus terrestris, Tephrosia purpurea.*

## b) Rabi Weeds

*Andropogon annulatus, Anagallis arvensis, Avena ludoviciana, Achyanthus aspera, Acanthosperum hispidum, Borreria hispida, Boerhavia diffusa, Cirsium arvense, Carthamus oxyacantha, Commelina benghalensis, Cleome viscosa, Daucus carota, Digera arvensis, Eleusine indica, Fimbristylis sp., Fumaria parviflora, Grosophora sp. Gynandropsis pentaphylla, Galinsoga parviflora, Heliotropium eichwaldi, Hibiscus micranthus, Indigofera* sp., *Isachne* sp., *Imperata cyalindrica, Leucas aspera, Lathyrus* sp., *Lagascea mollis, Ludwigia parviflora, Mimosa pudica, Nicotiana* sp., *Oxalis corniculata, Oldenlandia diffusa, Portulaca quadrifida, Physalis minima, Polygonum barbatum, Phyllanthus niruri, Phalaris minor, Panicum repens, Rumex acetosella, Setaria glauca Trianthema portulacastrum, Tridax procumbens, Vicia hirsuta, Vernonia cinerea, Vicia sativa, Wahlenbergia sp., Xanthium strumarium.*

The information collected from weed scientists all over the country was used to identify top ten weeds in dryland areas in *Kharif* and *rabi* seasons (Table 1).

Table 1 : Top ten weeds in dryland areas in India

| Kharif | Rabi |
|---|---|
| *Cyperus rotundus* | *Chenopodium album* |
| *Cynodon dactylon* | *Asphodelus tenufolius* |
| *Digitaria* sp. | *Cynodon dactylon* |
| *Ageratum conyzoides* | *Cyperus rotundus* |
| *Euphorbia sp.* | *Ageratum conyzoides* |
| *Elusine indica* | *Euphorbia* sp. |
| *Echinochloa* sp. | *Melilotus* sp. |
| *Cleome* sp. | *Convolvulus arvensis* |
| *Trianthema* sp. | *Argemone mexicana* |
| *Digera arvensis* | *Digitaria sp.* |

## CROP LOSSES

The crop losses due to weeds vary with several factors including crop cultivar and management practices, soil and climatic conditions etc. The crop losses caused by weeds have been determined for many dryland crops and are summarised in Table 2.

Table 2 : Crop yield losses due to weeds under dryland conditions

| Crop | % yield loss | Location | Reference |
|---|---|---|---|
| Groundnut | 30–50 | Punjab | Dalal (1967) |
| | 78 | Bangalore | Sanghi & Korwar (1987) |
| Pearlmillet | 25–50 | | Misra & Kumar (1962) |
| | 55 | | anghi & Korwar (1987) |
| | 81 | Akola | Sanghi & Korwar (1987) |
| Pigeonpea | 40 | Kanpur, Ali (1985) | |
| Fingermillet | 58 | Bangalore | Hegde & Seetharaman (1985) |
| Mothbean | 81 | Almora | Singh *et al.* (1990) |
| | 75 | Jodhpur | Henry & Singh (1985) |
| Chickpea | 30–50 | – | Panwar & Pandey (1977) |
| | 90 | Almora | Singh *et al.* (1990) |
| Upland Rice | 79 | Dehradun | Sanghi & Korwar (1987) |
| | 37 | Dehradun | Sanghi & Korwar (1987) |
| | 39 | Rewa | Sanghi & Korwar (1987) |
| Maize | 62 | Dehradun | Sanghi & Korwar (1987) |
| | 57 | Almora | Singh *et al.* (1990) |
| Sorghum | 60 | Hyderabad | Sanghi & Korwar (1987) |
| Soybean | 50 | Dehradun | Sanghi & Korwar (1987) |
| | 51 | Almora | Sanghi & Korwar (1987) |

Weeds and crop seeds germinate almost simultaneously and they start competing with each other for water, nutrients and space. If there is incessant rain during initial stages not only that the mechanical control measures are not possible but also it provides ideal conditions for weed emergence over an extended period of time. While, the moisture of surface soil is critical for weed emergence, the relative growth of crop plants - its population, ground cover etc. would decide about establishment and competitive ability of weed seedlings that have emerged. Hence there will be great variation in crop losses due to weeds from year to year and from place to place. The crop yield losses reported here (Table 2) have been normally worked out in experimental fields by recording the yield from weed free and weed infested treatments.

But rarely farmers allow weed growth which is completely unchecked. Unfortunately, reports regarding crop yield losses under farmers weeding practices are rare to find in the literature. Carson (1987) reported a grain yield loss of 27-36% by following farmers weeding practices in pearlmillet in Zambia. He also observed greater yield losses in drought year (15-67%) as compared to yield losses in favourable season (21- 35%).

Table 3 : Critical period for crop weed competition for crops grown in dryland areas.

| Crops | Critacal period | Location | Reference |
|---|---|---|---|
| Upland rice | 0–20 days | – | Sharma *et al.* (1977) |
| Pearlmillet | 30–45 days | Arid zone area | Misra & Kumar (1962) |
| Maize | 2–4 weeks | – | Sanghi (1985) |
| Fingermillet | 21–30 days | Bangalore | Hegde & Seetharaman (1985) |
| Pigeonpea | 7–8 weeks | Kanpur | Ali (1985) |
| Mothbean | 0–20 days | Jodhpur | Henry & Singh (1985) |
| Chickpea | 30–60 days | Hyderabad | Balasubramanian & Venkateswarlu (1985) |
| Soybean | 0–30 days | Indore, M.P. | Garg (1985) |
| Castor | 30–60 days | Gujarat | Patel (1985) |
| Cotton | 60–70 days | Akola, Maharashtra | Bhatwedekar (1985) |
| Moong | 0–4 weeks | – | Utomo (1987) |
| Groundnut | 15–60 days | – | Hill & Santelmann (1969) |
| | | | Bhan *et al.* (1971) |
| | | Punjab | Gill *et al.* (1971) |
| | | Jabalpur | Rathi *et al.* (1986) |
| Sunflower | 0–45 days | Bangalore | Krishna Gowda (1985) |
| Sorghum | 0–30 days | – | Burnside & Wick (1967) |
| Sesamum | 15–20 days | Jabalpur, M.P. | Sharma (1985) |

## CRITICAL PERIOD OF WEED COMPETITION

It is well established that the crop is most susceptible during its early stages of growth. However, the duration of critical period vary depending upon the crop, crop cultivar, duration, seed rate, spacing, planting geometry, growth habit etc. The critical stages of weed competition in different crops have been summarised in Table 3. Both weeds and crop plants germinate simultaneously and weeds very often out number crop plants. Early removal of weeds is very essential in order to place competition in favour of crop plants. Otherwise, weeds seize the opportunity, and cause irreparable damage to crop plants.

Crop or crop cultivar with early vigour and fast canopy cover is liable to suffer less from weed infestation. Short duration crops are likely to suffer more as the weed cycle is also short and the peak growth of crop will normally coincide with the peak growth of weeds too. In long duration crops, however, the crop has a chance to recover once the peak growth of weeds is over. Similarly, in wider spaced crops, intercropping has a distinct advantage in the sense that the vacant space between rows will not only be utilized properly but the technique will discourage weed growth also. Though weeding is still be required, contrary to claims made (Ali, 1988), the labour requirement and frequency of weeding are bound to be substantially reduced.

### Competition for Water

Plants differ in their ability to exploit available water supplies primarily due to variation in root system. Most severe competition between weeds and crops is centred around the supply of available soil moisture. Competition for water begins as soon as the root system of one plant invades a feeding area of another, and usually takes place long before tops are developed sufficiently to exert serious competition for light. Therefore, in dry climate roots actually decide the success or failure in competition between species otherwise equally adopted to a region.

Aldrich (1984) suggested several factors responsible for moisture competition. According to him both time and space are important aspects in competition for water. These dimensions are significant for weed-crop situations. There is a time immediately after emergence when the limited extent of root systems make competition improbable. The degree of competition for water between a crop and a weed is determined primarily by the relative root volume occupied by each. The density of weed infestation necessary to cause a given degree of competition for moisture varies from weed to weed.

In general, for producing equal dry matter, weed transpires more water than do most of our crop plants. The transpiration coefficients of major weeds of sorghum, *Amaranthus viridis, Cynodon dactylon, Digitaria sanguinalis* and *Echinochloa colonum* were 338, 813, 696 and 674, respectively as against 394 in case of sorghum (Karnitkar *et al.* 1960).

Wiese and Vandiver (1970) demonstrated the ability of crops and weeds to compete for soil moisture. Corn, sorghum, *Echinochloa crusgalli, Xanthium pensylvanicum* and *Digitaria sangunalis* achieved maximum growth in wet soils. Dry soils greatly reduced the growth of these species. *Kochia scoporia, Salsola kali, Schedonnardus paniculatus,* and *Solanum rostratum* produced less growth under wet conditions, but were unhampered in dry soil. In fact, *S. kali* doubled its dry weight in dry soils. Species that achieved greater growth in wet soil were most adversely affected by competition and water stress. Plant that produced little growth in wet soil actively competed in dry soil.

Blackshaw and Dekker (1988) studied the interference among rape, wild mustard (*Sinapis arvensis*) and lamb's quarter (*Chenopodium album*) in replacement series experiment under field conditions. Wild mustard was the strongest competitor for water followed by rapeseed and lamb's quarter. Similarly, greater competitive effect of red root pigweed and robust foxtail than that of soybean has been reported under water stress condition (Orwick and Schreibert, 1979). Kondap and Bhathkal (1981) observed that weeds absorbed more moisture than hybrid sorghum and affected the yield up to 24 to 68% as compared to weed free plots of sorghum. The evapotranspiration loss from 0-45 cm layer was 80 cm in unweeded sorghum as compared to 53 cm from weed free plots (Hosmani, 1976). Griffin *et al.* (1989) found soybean to be more competitive than the florida beggar weed (*Desmodium tortusum*), but under water stress conditions the weed was more competitive than soybean. However, Mortensen and Coble (1989) observed that the reduction in the growth of common cocklebur (*Xanthium strumarium*) was more than that of soybean in moisture stress conditions. They reported 29% reduction in yield of well watered soybean as compared to 12% in drought stressed crop. In Zambia, the pearlmillet suffered less yield loss due to weeds (29%) in favourable season as compared to 36% yield loss experienced in drought year (Carson, 1987).

Besides, photosynthetically more efficient $C_4$ plants have lower water requirement than $C_3$ plants. It is, therefore, interesting to note that out of ten major crop plants only maize is a $C_4$ plant, whereas eight out of 10 major weeds (as classified by Holm, 1969) belong to this group. In rice the upland direct sown crop is more vulnerable to weeds than the puddled transplanted crop. Besides other reasons, it is reported that the proportion of $C_4$ weeds increases with the increase in dryness of the soil.

## Competition for Light and Space

Light is the most important element of crop weed competition when moisture and nutrients are plentiful and weeds have an edge over crop plants in respect of growth in height (Gupta, 1984). The competition for light may commence very early in the crop season if a dense weed growth smothers the crop seedlings.

Bleardale (1959) emphasised that in dryland agriculture, in years of normal rainfall, the weed-crop competition was limited to nitrogen and light. By virtue of their number and faster growth, weeds often out-grow crop plants particularly in summer season. Hence early competition restrict crop growth and reduce plant stand substantially. This is true with even tall crops such as maize, sorghum and pearlmillet, which may tolerate late weed competition better but not early competition. Blackshaw and Dekker (1988) opined that competitive ability of *Sinapis arvensis, Chenopodium album* and *Brassica napus* appeared to depend upon their relative rates of development and competition for nutrients only played a minor role. Staniforth (1965), while working on soybean found that the yield depression was due to the competition for available soil water and light, because of shading effect of *Setaria.*

### Competition for Nutrients:

Weeds usually absorb mineral nutrients faster than many of the crop plants and accumulate them in their tissues in relatively larger amounts (Gupta, 1984).

The relative efficiency of weeds and crops in utilising available soil nutrients was determined at several dryland centres. And it was found that on an average, weeds extracted twice as much nitrogen, 25% more potassium and twice as much calcium as did crop plants. It is reported from Dehradun that about 20 kg nitrogen, 15 kg phosphorus and 10 kg potassium per hectare could be saved by application of alachlor to control weeds in pigeonpea, as compared to unweeded control (Sanghi & Korwar, 1987).

Kundra *et al.* (1990) reported that summer mungbean could utilise only 55.6, 10.2 and 49.1 kg/ha of N, P and K respectively while weeds removed 79.1, 19.8 and 79.1 kg/ha of those nutrients, respectively. Rajan and Sankaran (1974) reported 7-10 times more nutrient drain by *Trianthema* sp. than maize crop after 30 days of crop sowing.

## WEED MANAGEMENT

### Mechanical and Cultural Methods

In traditional cropping system, farmers wait until the onset of monsoon rains to begin pre-seeding tillage operations. Several cultivations may be necessary to control weeds which emerge with the availability of moisture. These cultivations may take several weeks resulting in exposure of the soil to serious water and wind erosion. The minimum tillage technique, on the contrary, is quite beneficial in conserving the soil. The major limitation being requirement of special attention on weeds. In developed countries, herbicides have become very handy for this purpose. However, the requirement of tillage has to be location-specific as pre-seeding tillage enables better infiltration of rain water.

Off season tillage in any form helps in reducing weed infestation (Table 4). It could be a post-harvest summer or even a pre-monsoon tillage. In post-harvest tillage, the late seed setting weeds are killed before they set seeds, thus reducing the weed seed input in the soil. In summer tillage, the flush of germinated/germinating weeds are removed, which reduces weed seed stock in the soil. In a 3-year study, off season tillage recorded yield increases ranging from 24-43% over control (Sanghi and Korwar, 1987).

Table 4 : Effect of tillage on weed infestation in red soils of Hyderabad.

| Treatment | Weed dry weight before seeding (kg/ha) | Weed dry weight before Ist interculture (kg/ha) |
|---|---|---|
| Post-harvest tillage | 0 | 2.24 |
| Summer tillage | 0 | 2.33 |
| Pre-monsoon tillage | 0 | 2.54 |
| Tractor tillage | 0 | 3.16 |
| Direct sowing | 2.78 | 6.16 |

Sanghi and Korwar (1987) also observed that the advantage of off- season tillage was more in dry season. This indicates the better effect of stored moisture through improved infiltration and the reduced loss of moisture through weeds, as a result of improved control through off season tillage.

Many dryland crops are widely spaced to facilitate inter-row cultivation. Weed growth between rows is normally controlled with bullock drawn blade harrows during 3 to 5 weeks after sowing. These operations are performed regardless of weed growth as they also break surface crust and create a dust mulch near the soil surface; which aids in moisture penetration during rainy periods and reduces evaporation during dry periods. Jain *et al.* (1985) reported that two intercultivations along with two hand weedings at 30 and 60 DAS gave the best weed control and produced the highest yield of seed cotton (1205 kg/ha).

However, it is always risky to rely entirely on intercultivation operations as weeds within the row can cause considerable losses in crop yield. Weed growth within the row is often neglected due to frequent rains or labour shortages during the critical weeding period. Weeds surviving in the row may reduce sorghum yields as much as weeds growing between the rows (Korwar and Friesen, 1985). In order to economise on the cost of herbicides, pre- emergence herbicides may be applied as a narrow band (1/4 to 1/3 the row space) over the row and weeds emerging between the rows can be controlled by interculture operations. One such study conducted at the Central Research Institute for Dryland Agriculture, Hyderabad indicated that weeding in the row either manually or chemically increased the grain yield of sorghum by over 50% despite 2 hoeings between the row (Table 5).

However, band application of pre-emergence herbicides to crop is difficult with traditional seeding methods, where a spine tooth harrow (*Jamboo*) is operated across rows immediately after sowing the crop, thus making it difficult to identify the seeded rows. Under such a situation, a post-emergence herbicide may be preferred (Table 6). Isoproturon applied at 3 WAS was comparable to pre-emergence application of atrazine in sorghum (Korwar and Friesen, 1985).

Table 5 : Effect of chemical and mechanical weeding on grain yield of sorghum

| Tratment* | Grain yield (g/m$^2$) | | |
|---|---|---|---|
| | No Atrazine | Atrazine 1.0 kg/ha blanket spray | Atrazine 0.33 kg/ha band application |
| No weeding | 202 (304)** | 290 (158) | 299 (142) |
| Weed free | 304 (35) | 320 (47) | 302 (27) |
| HW 2–3 WAS | 297 (34) | 296 (15) | 307 (42) |
| HW 3–4 WAS | 292 (33) | 296 (18) | 288 (18) |
| HW 4–5 WAS | 275 (21) | 307 (29) | 299 (20) |

* all treatments received two blade harrowings between the row, 3-5 WAS
**Weed dry wt. in g/m$^2$

Table 6: Weed control between and within the row on sorghum grain yield.

| Weed control | | Yield ($g/m^2$) |
|---|---|---|
| Between rows | Within rows | |
| None | None | 68 |
| 2 Hoeings | None | 123 |
| 2 Hoeings | One | 154 |
| 2 Hoeings | Atrazine band (0.33 Kg/ha) | 158 |
| 2 Hoeings | Atrazine broadcast (1.0 Kg/ha) | 160 |
| 2 Hoeings | Isoproturon band (0.33 Kg/ha) | 157 |
| 2 Hoeings | Isoproturon broadcast (1.0 Kg/ha) | 155 |

## Weed Control Through Agronomic Management

Various management practices starting from selection of crop, crop cultivar to spacing, planting geometry, intercropping, date of planting, fertiliser placement, crop rotation etc. have effect on weed growth. Although complete weed control through adoption of any single method is highly doubtful, combination of more methods would normally have complementary effect. As most of these methods do not involve extra cost, they are highly cost effective.

Each crop has its own characteristic weeds and they thrive well when the same crop is grown successively. Also, the associated weeds are adopted to the cycle of operations that follow with each crop. Any change in the crop or cropping would affect weeds. Double cropping is possible in some dryland areas and a shift from single to double cropping would be to the greater advantage of crops than weeds. Time of sowing is often not flexible in dryland areas but wherever possible, sowing can be manipulated to favour crop but not weeds.

Crops vary in their ability to suppress weeds. Among winter cereals, barley is reported to be more competitive than wheat or oat (Gupta, 1984). Higher tolerance of barley to weed competition is assigned to its ability to develop more extensive roots during the the initial stages of growth. Rao and Shetty (1976) observed that maize had the highest weed suppressing ability (92%) followed by pearlmillet (88%), sorghum (81%) and Setaria (77%). Amongst non-cereals, cowpea (88%) was more competitive than groundnut

(62%), castor (51%) or pigeonpea (54%). Similarly, *Cyperush rotundus* inflicted lower yield loss in maize and sorghum as compared to groundnut, soybean and grain legumes (Kondap *et al.*, 1980). In a 3-year study, high populations of *Xanthium pensylvanicum* were associated with soybean than with corn regardless of weed control measures (Walker and Buchanan, 1982).

Plant types with early vigour, rapid leaf area expansion are the ones which smother weeds better. Kulkarni *et al.* (1963) reported that groundnut 'TMV-2' (bunch type) suffered more due to weeds than 'TMV-3' (spreading type). Similarly, 'BH 8-18', a spreading type groundnut suppressed weed growth to a great extent as compared to 'Spanish improved' (Kavalappa *et al.*, 1988). Similarly, the weed growth in the compact genotype of pigeonpea was 37% higher than in spreading type (Rao and Shetty, 1976). The yield reduction in early maturing and short duration soybean 'Lee' was between 41 and 67%, as compared to 23-38% yield loss reported in tall, long duration cultivar 'Hardee' (McWhorter and Hartwig, 1972). In rice and wheat, modern cultivars have brought in increased problem of weeds mainly due to their shorter height. Shorter wheat cultivars have been reported to support about two times as much weed biomass (particularly of grass weeds) as that of traditional tall ones (Sandhu *et al.*, 1981). The yield loss was 25% in tall as compared to 55% in dwarf cultivar.

Higher plant population of the crop can put pressure on availability of space for weed growth from relatively early stages of crop growth. Although reports of better control of weeds with higher seed rate, narrow row spacing, cross sowing etc. particularly in wheat are available (Dhiman *et al.*, 1985; Prakash *et al.*, 1986), these practices seem to have less application in dryland areas because of input constraints. However, Burnside and Wicks (1969) observed that sorghum planted at 100 cm rows experienced 4, 12 and 18% yield loss when weeded at 3, 4 and 5 WAS respectively. But with 50 cm rows, higher grain yields were maintained even with increasing weed competition.

Intercropping is widely practised in dryland areas. Besides covering risk, intercropping enables efficient utilization of resources. In wide spaced crops, intercropping helps in better utilization of sunlight besides smothering the weeds. The planting of early maturing grain legumes such as mung bean (*Vigna radiata*(L.) Wilczek) or cowpea (*Vigna ungiculata* (L.) Walp.) as intercrops minimised weed infestation in sorghum, pigeonpea, pearlmillet and cotton (De, 1974; Rao and Shetty, 1977). In pigeonpea/sorghum system, cowpea was better than mung. In pearlmillet/groundnut system, the row arrangement of one row of pealmillet with three rows of groundnuts resulted in optimum weed suppression and maximum intercrop advantage. Ali (1988) observed that in rainfed pigeonpea, weed control can be overlooked if intercropped with short duration, close canopy forming and fast growing grain legumes such as cowpea, urd bean (*Vigna mungo*) and mung bean. He reported 20-45% reduction in weed growth in intercropped over sole pigeonpea.

## Weed Control Through Herbicides

In India, chemical weed control has not been widely accepted in dryland agriculture because of the availability of farm labour at low wages, the scarcity and high cost of commercially available herbicides and sophisticated technology of herbicide usage. The common adoption of mixed cropping, involving both dicotyledonous and monocotyledonous crops, is also a contributing factor as it is difficult to develop selective herbicides for mixed cropping systems. The use of herbicides for dryland crops may, however, become necessary under some specific situations. These include: when land is inaccessible for manual or mechanical weeding because of soil wetness; holdings are too large for manual weeding; a rise in cost of labour or a shortage of labour during peak periods of demand; in situation where weeds are not easily controlled by other methods etc.

Over the past many years, adequate studies have been made in the field of chemical weed control under dryland conditions. Table 7 gives some guidelines about the use of locally available herbicides in dryland crops.

Table 7 : Guide to herbicides use in dryland areas

| Crop | Herbicide | Dose (kg/ha) | Application | Remarks |
|---|---|---|---|---|
| Maize | Atrazine (Atrataf) | 1.0–1.5 | Pre-em or early post-em. | Apply to well prepared moist soil or when rain can be expected within a few days. |
| | Alachlor (Lasso) | 1.5–2.0 | Pre-em. | Mix with atrazine for broadspectrum weed control. |
| | 2,4–D | 1.0 | Post-em | Use as directed spray on maize over 30 cm high. |
| | Pendimethalin (stomp) | 1.5–2.0 | Pre-em | Mix with atrazine for broadspectrum weed control |
| Sorghum pearl millet | Atrazine | 0.5–1.0 | Pre-em or early post-em | |
| | 2,4–D | 0.5–1.0 | Post-em | Use as directed spray |
| Upland rice | Butachlor (Machete) | 1.0–2.0 | Pre-em | Pre-plant incorporation under dry conditions. |
| | 2,4–D | 0.5–1.0 | Post-em | After tillering and before shooting |
| | Pendimethalin | 1.0–1.5 | Pre-em | |
| | Thiobencarb (Saturn) | 1.0–2.0 | Pre-em | |
| Wheat | 2,4–D | 0.5–1.0 | Post-em | From 5 leaf stage to jointing or shooting. Some cultivars are susceptible. |

Table 7 contd...

| | | | | |
|---|---|---|---|---|
| | Isoproturon | 0.75-1.0 | Pre-em or Post-em | Require moisture for activity |
| | Pendi methalin | 1.0-1.5 | Pre-planting or pre-em | |
| Cotton | Diuron (Karmex) | 0.5–1.0 | Pre-em | Do not apply in coarsae sandy soil |
| | | | Post-em | Directed at lay by when cotton is atleast 30cm tall. |
| | Fluchloralin (Basalin) | 0.75–1.5 | Pre-planting | Incorporate into soil immediately after application |
| | Pendimethalin | 1.5–2.0 | Pre-em | Apply as pre-planting incorporated under dry conditions |
| Finger millet | 2,4–D | 0.5–0.75 | Post-em | |
| Pigeonpea | Alachlor | 1.0–1.5 | Pre-em | |
| Soybean | Fluchloralin | 0.75–1.0 | Pre-plant incorporated | |
| Urd bean | Pendimethalin | 1.0–1.5 | Pre-em | |
| Mung bean Cowpea groundnut | Fluajifop p-butyl (Fusilade) | 0.125–0.200 | Post-em | Contols grasses only |

The correct dose and time of herbicides treatment is, however dependent on several local factors including soil, climate, cropping sequence, management practices etc. Following are some of the points which need to be looked into while following chemical weed control under dryland conditions:

(1) Normally two or more crops are grown in inter or mixed cropping systems. Unlike other pesticides there is need to use different herbicides in different crops to control, some-

times, the same weeds. Therefore, due care must be taken to select a herbicide which is safe for both the crops. Fluchloralin and pendimethalin are safer in many crops, hence ideal for intercropping situations.

(2) Lowest doses of herbicides are required in soils which are light and which contain lower organic matter. This is essential to avoid crop damage, as unseasonal heavy rains may leach the herbicide to the crop seed zone.

(3) Because of low rainfall, relatively higher dose of herbicides than normally required under irrigated conditions, is required for weed kill in dryland areas. Besides, the herbicide persistence in soil is greater under dry conditions because of poor microbial and chemical degradation of the herbicide in the soil.

(4) For better performance, certain amount of moisture in the surface soil is necessary. Hence application of herbicides either on moist soil or when the rains are expected within 8-10 days of application is recommended.

(5) Incroporation of herbicides in the surface 3-4 cm soil layer becomes necessary under very dry conditions.

(6) One particular herbicide often may not be effective against all types of weeds, thus necessitating tank mixing of two or more herbicides. However, good understanding of the interactive effect of herbicides is necessary before the use of herbicide mixture.

(7) Although weed control through chemicals is as good as one or two manual weedings, with some herbicides, particularly at lower doses, a late manual weeding may become necessary to eliminate weeds which emerge late or which have escaped treatment. Very good results have been reported with herbicide + hand weeding or intercultivation in groundnut, (Malavia and Patel, 1986; Sanghi and Korwar, 1987), sunflower (Krishna Gowda *et al.*, 1985) and cotton (Jain *et al.*, 1985).

(8) Some differences in tolerance of cultivars to herbicides may exist, therefore, it is worth checking before treatment.

(9) Hormone type weed killers are highly phytotoxic on sensitive crops such as cotton. Care is, therefore, required to see that there is no drifting of the herbicide spray to the neighbouring fields where susceptible crops are grown.

(10) Herbicide may be applied as a narrow band over the row (as discussed before) and intercultivation done to check the weeds between the rows. This will save atleast 1/3 to 1/2 of the herbicide cost. Besides, there is less danger of contaminating the soil.

## SPECIAL WEED PROBLEMS IN DRYLAND CROPS

**Nutsedge** ***(Cyperus rotundus)*** **:** Nutsedge is a serious problem particularly in black soils. Its control has been studied at Hyderabad through the use of different herbicides. To date, the only herbicide showing promise is glyphosate at 2 kg/ha applied during *kharif* fallow after the sedge has attained a height of 10-15 cm (Sanghi and Korwar, 1987).

Split applications appear to improve the effectiveness of this herbicide. Whenever, the weed infestation is in patches, only the affected patches need be sprayed rather than a blanket application. This reduces the cost of treatment, which is essential if dryland farmer is to adopt the practice.

***Cynodon dactylon*** **:** *C. dactylon* is another problematic weed in the heavy soils. At Ranchi, various tillage practices have been compared for its control. Observations on regrowth showed that deep ploughing gave sufficient control of this weed and provided a competitive advantage to sorghum, pigeonpea and pearlmillet. The use of a cutivator, disc or the wooden plough were less effective (Sanghi & Korwar, 1987).

***Striga*** **:** The parasitic weed *Striga asiatica* in sorghum and pearlmillet is a serious problem in many areas, more so in red sandy loams, mixed soils or shallow black soils. The high yielding sorghum hybrids are particularly more susceptible to this weed. Yield loss have been observed to the extent of 100%. Studies at Hyderabad have shown that application of 2, 4-D @ 2 kg/ha at 4 WAS of sorghum reduced *Striga* emergence to a large extent and increased yields considerably (Sanghi and Korwar, 1987). The application of 2,4-D after *Striga* emergence (at flowering stage) was helpful in killing the emerged *Striga* plants but not as efficient in increasing sorghum yields as the former treatment. Almost similar results were reported by Yaduraju and Hosmani (1979).

As substitute to chemical control, alternate crops have been studied. Pearlmillet was not infested by the existing strain of *Striga* at Hyderabad. The other is intercropping of sorghum with pigeonpea. If sorghum is infested with *Striga,* the intercrop of pigeonpea, compensates for the reduction in yield.

Cultural method plays an important role in the control of *Striga.* Rotation of susceptible crops with certain crops (trap crops) has been reported to reduce the infestation substantially. Yaduraju and Hosmani (1979) observed that trap cropping with cowpea, groundnut, linseed and cotton recorded significantly lower number of *Striga* in the succeeding sorghum crops. Application of higher doses of N fertilizers also decreased the *Striga* infestation in sorghum considerably (Yaduraju *et al.,* 1979).

A novel way of controlling *Striga hermonthica* using the antitranspirant has been proposed recently (Prees *et al.,* 1989). The leaf temperature of *S. hermonthica* are substantially below those of the air - the difference may be as high as 7° C at 40° C. This results from high rates of transpiration and the consequent cooling of the leaf. Prees *et al.* (1989) reported that application of an antitranspirant increased the leaf temperature of *Striga* by 15-20% due to reduced transpiration (40% less) and stomatal conductance (57% less) as compared to unsprayed one. *S. hermonthica* is very sensitive to high temperature and application of an antitranspirant under field conditions in the Sudan increased the grain and straw yield of sorghum by factors 3.4 and 2.6 respectively (Table 8).

Table 8 : Effect of spraying *S. hermonthica* with antitranspirant on yield of parasite and its host sorghum

| | Unsprayed | Sprayed |
|---|---|---|
| *S. hermonthica* dry wt. (kg/ha) | 1697 | 1057 |
| Sorghum grain yield (kg/ha) | 88 | 298 |
| Sorghum straw yield (kg/ha) | 617 | 1575 |

*Pluchea lanceolata* : *P. lanceolata* is another troublesome perennial weed in light soils of northern India. Being deep rooted, this weed is particularly harmful to dryland crops as it exhausts underground stored moisture. Experiments carried out at IARI, New Delhi by Yaduraju and Ahuja (1989) revealed the excellent activity of clopyralid 0.15 kg/ha applied either once or twice (Table 9). Repeated application of 2,4-D or fluroxypyr at 0.4 kg/ha also resulted in significant reduction of *Pluchea* biomass. The regrowth obtained following the cutting of treated plants was also significantly less in these treatments.

Table 9 : Effect of weed control treatments on the growth of *Pluchea*

| Treatments | Application | Dry wt. $g/m^2$ | | Regrowth (12 weeks after cutting) |
|---|---|---|---|---|
| | | 6 WAT | 18 WAT | |
| 2, 4-D 0.4 kg/ha | Single | 25 | 90 | 80 |
| 2, 4-D 0.4 kg/ha | Double | 23 | 60 | 43 |
| Fluroxypr 0.4 kg/ha | Single | 17 | 110 | 56 |
| Fluroxypr 0.4 kg/ha | Double | 24 | 91 | 68 |
| Clopyralid 0.15 kg/ha | Single | 29 | 76 | 39 |
| Clopyralid 0.15 kg/ha | Double | 25 | 43 | 13 |
| Imazethapyr 0.15 kg/ha | Single | 22 | 129 | 94 |
| Imazethapyr 0.15 kg/ha | Double | 24 | 148 | 89 |
| Mechanical cutting | – | 5 | 120 | 49 |
| Untreated control | – | 26 | 220 | 144 |
| CD at 5% | – | NS | NS | 66 |

Second application made after 15 days of first application

Also it was observed in another experiment that addition of 0.5% sugar to the spray solution increased the efficiency of 2, 4-D.

*Convolvulus arvensis* : Field bindweed, as it is commonly called, is a prostrate or climbing perennial with an extensive root system. It is most troublesome in light soils and during winter season. The propagation is by seeds but its extensive root system and its ability to regrow following destruction makes control difficult. Shoots of *C. arvensis* are readily killed by the phenoxyalkanoic acids (2, 4-D etc.) as well as by mixtures containing dicamba. Post-emergence applications of fluroxypyr at 0.4 or 0.6 kg/ha applied once or repeat application at 0.4 kg/ha were very promising for the control of *C. arvensis* (Ahuja and Yaduraju, unpublished). 2, 4-D at similar doses was also equally effective but not clopyrolid, isoproturon or paraquat. In U.K., post emergence application of oxadiazone at 2 kg/ha has been found to be highly effective against field bindweed in orchards, vineyards and woody horticultural crops (Boyall, 1983). Control of most weeds is best when they are young but it has been shown that *Convolvulus* is an exception and is most susceptible when it is well developed.

Although, susceptible to cereal herbicides containing phenoxyalkanoic herbicides or dicamba, bind weed often causes problems in cereals because it emerges too late to be controlled by them. It climbs upto the stems of the cereals where it can impede harvesting and sets seed, thereby becoming an increasing problem in subsequent seasons. In U.K., pre-harvest application of glyphosate is reported to give very good control of this weed in cereal crops (O'Keeffe, 1980) either 1.44 or 2.16 kg/ha having given 95% control one year after harvest.

## REFERENCES

Aldrich, R.J. (1984) Weed Crop Ecology - Principles in Weed Management. Breton Publishers PP: 189-274.

Ali, M. (1985) Pigeonpea. *In*: Efficient Management of Dryland Crops (eds.) V. Balasubramanian and J. Venkateswarlu. pp. 143.

Ali, M. (1988) Weed suppressing ability and productivity of short duration legumes intercropped with pigeonpea under rainfed conditions. Tropical Pest Management, 34(4): 384-387.

Balasubramanian, V. and Venkateswarlu, J. (1985) Chickpea *In* : Efficient Management of Dryland Crops. (eds.) V. Balasubramanian and J. Venkateswarlu, pp. 209.

Bhan, V.M., Megh Singh and Maurya, R.A. (1971) Crop-weed competition studies in groundnut. Indian J. Weed Sci. 3(1): 32- 36.

Bhatawadekar, P.U. (1985) Cotton *In* : Efficient Management of Dryland Crops (eds.) V. Balasubramanian and J. Venkatwswarlu, pp. 293.

Blackshaw, R.E. and Dekker, J. (1988) Interference among *Sinapis arvensis*, *Chenopodium album* and *Brassica napus*. I. Yield response and interference for nutrients and water. Phyto- protection; 69: 105-120.

Boyall, L.A. (1983) The control of perennial weeds. *In* : Recent Advances in Weed Research (eds) W.W. Fletcher, CAB, London. pp. 141-170.

Burnside, O.C. and Wick, G.A. (1967) The effect of weed removal treatments on sorghum growth. Weed Sci., 15: 204-207.

Burnside, O.C. and Wick, G.A. (1969) Influence of weed competition on sorghum growth. Weed Sci. 17: 332-334.

Carson, A.G. (1987) Improving weed management in the draft animal based production of pearlmillet in the Gambia. Tropical Pest Management, 33(4): 359-365.

Carson, A.G. (1989) Effect of intercropping sorghum and groundnut on density of *Striga hermonthica* in the Gambia. Tropical Pest Management, 35 (2): 130-132.

Dalal, J.L., Gill, H.S. and Saini, J.S. (1967) Effect of sowing, weeding - cum - manuring on the yield of rainfed groundnut in Punjab. Indian J.Agron. 12(3): 309-313.

De, R. (1974) Prospects for crop intensification and diversification in rainfed areas in Southeast Asia FAO/UNDP International expert consultation on the use of Improved Technology for Production in Rainfed Areas of Tropical Asia 1-7, December 1974 Khan Karan (Thailand)

Deshpande, R.M.; Thakare, V.S., Patil, M.B. and Gomase, B.P. (1987) Effect of weed removal at different times on the yield of rainfed American cotton. PKV Research Journal, 11(2): 174-175.

Dhiman, S.D.; Rammohan, D.S. and Sharma, H.C. (1985) Studies on cultural methods of weed control in wheat. Indian J. Agron., 30(1): 10-14.

Garg, K.P. (1985) Soybean. *In* Efficient Management of Dryland Crops (eds). V. Balasubramanian and J. Venkateswarlu, pp.259.

Gill, B.S.; Sarma, J.S.; Brar, H.S. and Mehra, S.P. (1986) Studies on the critical period of weed competition in groundnut. J. Res. Punjab Agril. Univ., 23(3): 394-97.

Griffin, B.S., Shilling, D.G., Bennett, J.M. and Currey, W.L. (1989) The influence of water stress on the physiology and competition of soybean and florida beggar weed (*Desmodium tortuosum*) Weed Sci.,

37(4): 544-551.

Gupta, O.P. (1984) Scientific Weed Management. Today and Tomorrow Printers & Publishers, New Delhi.

Hegde, B.R. and Seetharaman, A. (1985) Finger millet and other minor millets. *In*: Efficient Management of Dryland Crops (eds.) V. Balasubramanian and J. Venkateswarlu, pp. 79.

Henry, A. and Singh, R.P. (1985) Mothbean. *In* Efficient Management of Dryland Crops PP. 199.

Hill, L.V. and Santelmann, P.W. (1969) Competitive effects of annual weeds on Spanish peanuts. Weed Sci. 17(1) : 1-2.

Holm, L. (1969) Weed problems in developing countries. Weed Sci. 17 : 113-118.

Hosman M.M. (1976) Evapotranspiration losses in sorghum and maize infested by weeds. Proc. Indian Soc. Weed Sci. Conf., Hyderabad.

Jain, H.C., Jain, S.C. and Deshmukh, S.C. (1985) Cultural and chemical weed control in cotton (*Gossypium hirsutum* (L.)) in Nimar Region of Madhya Pradesh. Indian J. Weed Sci. 17(4): 1- 8.

Karnitkar, N.V., Sirur, S.S. and Gokhale, D.H. (1960) Dry farming in India. ICAR Pub. 2nd Ed., New Delhi, pp. 241.

Kavalappa, B.N.; Ramachandra Prasad, T.V., Dwarkanath N. and Krishnamurthy, K.(1988) Nutrient uptake in groundnut and weed as influenced by weed control treatment. Current Research, University of Agr, Sciences, Banglore, 17 (4): 45-46.

Kondap, S.M., Reddy, M.R. and Rama Krishna, K. (1980) Studies on yield reduction *Cyperus rotundus* in different crops. Abst. Conf. Indian Soc. Weed Sci. 1980. P.25.

Kondap, S.M. and Bathkal, B.G. (1981) Crop weed competition studies in sorghum under different management practices. *In* : Proc. 8th Asian Pacific Weed Sci. Soc. Conf. Vol.II, pp 117-122, Banglore, India.

Korwar, G.R. and Friesen, G.H. (1984) Control of *Strigaasiatica* in sorghum. Tropical Pest Management. 30(1): 14-17.

Korwar, G.R. and Friesen, G.H. (1985) In-the-row weed management in sorghum. Tropical Pest Management. 31(1): 324-25.

Krishna Gowda, K.T., Muniyappa, T.V. and Venkataramu, M.N. (1985). Weed control in sunflower through mechanical, chemical and cultural methods. Indian J.Weed Sci. 17 (4) : 49-51.

Kulkarni, L.G., Verma, S.S. and Achoto Rao, K. (1963) Studies on weeding and intercultures in relation to weed control in the yield of groundnut. Indian Oilseed J.I (1): 126-129.

Malavia, D.D. and Patel, J.C. (1986) Studies on integrated weed control in rainfed groundnut (*Arachis hypogea*) Indian J. Agron. 31(4): 338-342.

McWhorter, C.G. and Hartwig, E.E. (1972) Competition of Jhonson grass and cocklebur with six soybean varieties. Weed Sci. 20(1) : 56-59.

Misra, D.K. and Kumar, V. (1962) Response of *Pennisetum typhasides* to weeding in arid zone farming. Indian J. Agron. 6 : 260-269.

Martensen, D.A. and Coble, H.D. (1989) The influence of soil water content on common cocklebur (*Xanthium strumarium*) interference in groundnut. Weed Sci. 37: 76-83.

O'Keeffe, M.G. (1980) The control of *Agropyron repens* and broad leaved weeds pre-harvest of wheat and barley with the isopropylamine salt of glyphosate. Proc. British Weed Control Conf. 1: 53-60.

Orwick, P.L. and Schreiber, N.M. (1979) Interference of redroot pig weed (*Amaranthus retroflexus*) and robust foxtail (*Setaria viridis var robusta alba* or var. robusta- purpurea) in soybean (*Glycine max*). Weed Sci. 27(6): 665-674.

Panwar, K.S. and Pandey, K.(1977) Weed control studies in Bengalgram. Indian J.Agron. 22: 257-259.

Patel, R.R. (1985) Castor *In* : Efficient Management of Dryland Crops. PP. 293.

Prakash, A., Sharma, K.C. and Mishra, R.D. (1986) Control of *Phalaris minor* and wild oats in wheat by cultural and chemical methods. Indian J. Agron. 31(4): 411-413.

Prees, M.C., Nour, J.J., Bebawi, F.F. and Stewart, G.R. (1989) Antitranspirant-induced heat stress in the parasitic plant *Striga hermonthica* - A novel method of control. J. Expt. Bot. 40 (214): 585-591.

Rajan, A.V. and Sankaran, S. (1974) Studies on crop weed competition for nutrients and its effects on grain yield of maize (Var. Ganga-5). Madras Agric. J. 61(4): 413-416.

Rangaswami, G. (1983) Biodegradation of pesticides. Pesticide Information 9(3): 55-63.

Rao, M.R. and Shetty, S.V.R. (1976) Some biological aspects of intercropping systems on crop weed balance. Indian J.Weed Sci. 8: 32-43.

Rathi, G.S., Sharma, R.S. and Dubey, M.P. (1986) Studies on integrated weed control in rainfed groundnut. Indian J. Weed Sci. 18 (4): 220-225.

Sandhu, K.S.; Kolar, J.S. and Singh, G. (1981) Studies on cultural control of, wildoats (*Avena ludoviciana* Durr.) a grassy weed in wheat. Indian J. Weed Sci. 13: 146-149.

Sanghi, N.K. (1985) Maize, *In* Efficient Management of Dryland Crops. PP. 63.

Sharma, H.C. Singh, H.B. and Friesen, G.H. (1977) Competition from weeds and their control in direct seeded rice. Weed Res. 17: 103-108.

Sharma, S.M. (1985) Sesamum, In. Efficient management of dryland crops.: 245.

Shetty, S.V.R. and Rao, M.R. (1977) Weed management studies in pigeonpea based intercropping. Presented at the 6th Asian Pacific Weed Science Society Conference, 11-17, July, 1977. Jakarta Indonesia, 25 pp.

Staniforth, D.W. (1965) Competitive effect of three foxtail species on soybeans. Weeds 13(3): 191-193.

Utomo, I.H. (1987) Critical period of mung bean (*Phaseolus radiatus* L.) to weed competition. Biopopia (1988/89) No. 2, 8-11. Weed Abstr. 38(8): 283.

Walker, R.H. and Buchanan, G.A. (1982) Crop manipulation in integrated weed management systems. Weed Sci. 30 (Suppl.1): 17-24.

Wiese, A.F. and Vandiver, C.W. (1970) Soil moisture effects on competitive ability of weeds. Weed Sci. 18(4): 518-519.

Yaduraju, N.T. and Ahuja, K.N. (1989) Chemical control of *Pluchea lanceolata* in non-cropped area. *Indian J. Weed Sci. 21* : 51.

Yaduraju, N.T. and Hosmani, M.M. (1979) *Striga asiatica* control in sorghum, PANS 25(2): 163-167.

Yaduraju, N.T., Hosmani, M.M. and Prabhakara Setty, T.K. (1979) Effect of time and dose of nitrogen application and *Striga asiatica* incidence in sorghum. Proc. Second Symposium Parasitic Weeds: 285-288.

Dryland Resources & Technology, 7 (1992)
Editors : L.L. SOMANI, R.L. SHYAMPURA
& M.C. MUNDRA


# Rainfed Maize - Recent Advances in Production and Protection Technology

*G.L. Jain*
Associate Director Research,
Agricultural Research Station,
Raj. Ag. Univ., University Campus,
UDAIPUR–313 001

## ABSTRACT

Maize in India is mainly planted in *Kharif* season as a rainfed crop. Since the inception of All India Coordinated Maize Improvement Project in the year 1956-57, a large amount of technology has been generated through the project as well as various Agricultural Universities and ICAR research institutes. Most of the research work thus generated has been published. To bring this technology at one place an attempt has been made so that the same can be disseminated to various agricultural agencies who are concerned with the production of maize in India.

The agro-technology has been divided in various sections like superior cultivars and different dates of planting with the onset of monsoon under various agroclimatic zones, plant population and its pattern, planting methods, weed control and inter-tillage, application and management of fertilizers including bio- fertilizers for sole as well as intercrops, rain water management, intercropping techniques to boost the production based on maximum utilization of available natural resources like water, air, sunshine and nutrients, role of plant growht regulators, protection against pests and diseases, etc.

Each agro-techiques has its own role to increase the production singly but when combined with other techniques it interacts for positive action. For example, weed control alone from 4 week to 15 week age of maize crop increase the production from 96.5 to 73.3 percent. Thus, if the improved agro-techniques are adopted, there is no doubt that production can be increased at least upto 200 percent.

## INTRODUCTION

Maize in India is mainly planted in *kharif*. As a rainfed crop, its cultivation in India extends, under diverse conditions of rainfall, from the hot arid plains of Rajasthan and Gujarat to the wet hills of Assam and Bengal. The rainfall in the maize growing areas ranges between 380-635 mm in the North-West plains of Rajasthan to about 2030 mm in the hills of Darjeeling and Kalimpong. Well distributed rainfall in adequate amount represents an ideal condition for maize production, as most of it in India is grown as rainfed. The States of Uttar Pradesh, Bihar, Andhra Pradesh, Rajasthan, Punjab, etc. wehre it is grown in *kharif* account for about 75% of the area and production of this cereal in the country.

In rainfed maize, improved cultural practices can bring about 200% increase in yield. This fact has been brought out by agronomic trials conducted over a wide range of agro-climatic conditions in the country.

## AGRONOMIC SUPERIORITY OF MAIZE CULTIVARS

Improvement of the crop on a nation wide basis was undertaken in 1957 under the auspices of All India Co-ordinated Maize Improvement Project. The details of more than 50 prominent cultivars released which are prevelent for general cultivations are enumerated.

### [A] Hybrids

1. **Deccan Hybrid** : Grain bold, yellow and semi-flint, Maturity 100-110 days, recommended for A.P., M.P., Gujarat, Karnataka and Maharashtra. Excellent husk cover. Highly resistant to leaf blight and downy mildew.

2. **Sartaj** : Yellow orange flint, Maturity 95-100 days. Recommended for across the country.

3. **Ganga safed 2** : Grain-Medium, white, semi-flint with purple grain at low frequency, medium maturity (100-110

days). Recommended for all states except hilly areas and Karnataka state. Resistant to bacterial and pythium stalk rot. Tolerate salinity more than others. Does well under rainfed condition.

4. **Hi-starch** : Grain-bold, white-dent, high starch recovery, medium maturity. Adapted in Bihar and UP. Does also well in *Rabi*.

5. **Him. 123** : Grain-bold yellow and semi-flint, Maturity-100 to 110 days. Recommended for cultivation upto: 1, 800 metres (H.P., J&K, Assam, Nagaland and hills of U.P.). Excellent huskcover, Highly resistant to downy mildew, leaf blight and rust. Susceptible to stalk rot under water-logging.

6. **Ganga-5** : Grain-bold, yellow and semi-flint. Maturity - 95 to 110 days. Medium statured and recommended for all states except Karnataka. Resistant to leaf blight, brown strips, downy mildew, stem borer and drought. It has also shown promise in a number of Asian countries.

7. **Ganga-4** : Grain-bold, white and flint. Recommended for M.P., Bihar and U.P. (Tarai belt) and Rajasthan. No purple grains. More resistant to downy mildew. Flowers 3-5 days later and yields 13% more than Ganga 2.

8. **Ganga-9 (Ex Hy 2380)** : Grain-yellow, semi-flint, medium maturity (100-110 days). Recommended for Assam, Bihar, HP, J&K, Manipur, Meghalaya, Nagaland, Sikkim, UP, WB, Arunachal Pradesh and Delhi.

9. **Himalayan-128** : Grain yellow, semi-flint, full season (100-110 days). Recommended for cultivation in the Kumaon Hills of H.P., HP and Sikkim.

10. **Deccan- 103** : Released in 1981. A medium maturing (100-110 days) yellow seeded semi-flint hybrid resistant to leaf blight, sorghum downy mildew and stalk rot. Recommended across the country.

## [B] Composites

### (a) Long Duration :

1. **NLD :** White, semi-flint, Maturity 100-110 days. Adapted for Meghalaya, Nagaland, Sikkim, WB, Arunachal Pradesh and Mizoram.
2. **COI :** Yellow semi-flint. Maturity 100-110 days. Recommended for TN.
3. **Kisan :** Grain-medium yellow, semi-flint, Maturity 95- to 110 days. Marked resistant to downy mildew and leaf blight. Adapted across the country except AP, Gujarat, HP, Karnataka, MP, Maharashtra, Orissa and TN.
4. **Vikram :** Grain-medium yellow, orange, semi-flint, Maturity 90 to 100 days. Recommended for across the country except AP, Assam, HP, JK, Karnataka, Maharashtra, Sikkim, TN, WB, Arunachal Pradesh, & Mizoram. Considerable resistance to drought and downy mildew.
5. **Vijay :** Grain-medium yellow flint to semi-flint, Maturity 95 to 100 days. Wide adaptability in Indian subcontinent except Karnataka. Considerable resistance to foliar diseases

### (b) Medium Maturity

1. **Ageti-76 :** Earlier in flowering by 4-5 days than Vijay. 90-95% yield of Vijay. Has reasonable level of tolerance to most of the diseases and pests of local importance. Grain-orange semi-flint. Maturity 90-95 days. Recommended for across the country except Karnataka, Raj., TN, and Hilly areas of UP.
2. **Tarun :** Medim early maturing composite, yellow semi-flint grains. Considerable tolerance to brown stripe downy mildew and other foliar diseases. Maturity 85-90 days. Recommended for Bihar, Haryana, Punjab, UP and Delhi.
3. **Prabhat :** Grain type yellow semi flint. Maturity 85-95 days. Area of adaptation : Bihar, Gujarat, Karnataka, MP, Maharashtra, Punjab, TN and UP.

4. **Navjot** : Grain type : orange yellow semi-flint. Maturity 85-95 days. Recomended for across the country.

5. **Mansar** : Orange yellow semi-flint. Maturity 85-95 days. Recommended for HP, JK and Hill regions of UP.

6. **Harsh** : Orange yellow semi flint. Maturity 95-100 days. Recommended for AP, Karnataka, MP and Maharshtra.

(c) **Early Maturity**

1. **Hunius** : Grain type-orange semi-flint. Maturity 80-85 days. Recommended for Assam, Gujarat, HP, JK, Maharasthra, Sikkim, UP, Delhi and Mizoram.

2. **Diara - 3** : Grain orange yellow semi-flint. Maturity 75-80 days. Recommended for across the country except Karnataka.

3. **D-765** : Grain orange semi-flint. Maturity 75-80 days. Recommended for across the country except Karnataka.

4. **MCU 508** : Grain yellow semi-flint. Maturity 75-80 days. Recommended for across the country except Haryana, Karnataka, Punjab and Sikkim.

5. **Arun** : Grain orange yellow semi-flint. Maturity 80-85 days. Recommended for Punjab, Rajasthan, Delhi, Haryana, UP and Diara region of Bihar.

6. **Renuka** : Grain yellow semi-flint. Maturity 75-80 days. Recommended for Karnataka.

7. **VL Makka-88** : Grain yellow semi-flint. Maturity 80-85 days. Recommended for UP and HP hills.

8. **Kiran** : Grain type orange flint. Maturity 80-85 days. Recommended across the country.

## STATE RELEASE UNDER CULTIVATION

### (A) Hybrids (i) U.P.

1. **Ganga-7** : Yellow bold-seeded medium maturity, wide adaptation. Comparable to Ganga-5 in yield. High degree of resistance to brown stripe downy mildew and better

standability.

2. **VL - 41 :** Adapted to hills, yellow flint seeded, short duration, 80-85 days and wide adaptability.

3. **VL-16 :** Grain orange yellow semi-flint, maturity above 80 days and suitable for hills ranging 600-2400 meters a.s.1.

4. **VL Amber Pop :** Grain yellow pop, maturity 100-105 days on lower hills and 115-120 days on higher hills.

5. **Naveen :** Grain type yellow flint, maturity 90-95 days and recommended for Plains.

6. **Kanchan** : Grain orange semi-flint, maturity 75-80 days and recommended for UP.

## (B) Composites

### (I) PUNJAB

1. Makki safed 1 : Medium early maturing, very promising but susceptible to brown stripe downy mildew.
2. Pratap
3. Navjot : Yellow grain and maturity 85 days

### (II) ANDHRYA PRADESH

1. **Amber pop;** Promising composite, yellow pop and maturity 80-85 days.
2. **Rohini :** Grain orange semi-flint and maturity 90-100 days.

### (III) BIHAR

1. **Diara composite :** Very early maturing cultivar (75- 80 days) with yellow flint grains. Recommended particularly for Diara areas of Bihar. Does well at low level of N.
2. **Rajendra Makka 1 :** Early maturing composite of Diara maturitty, yellow semi-flint grains and good performance in Diara areas.
3. **Suwan**–1 (Composite) : Grain yellow semi-flint, maturity 95-100 days. Recommended for plains.
4. **Laxmi.**

### (IV) UTTAR PRADESH

1. Shweta
2. UL Umber pop
3. Navin

4. Kanchan

**(V) WEST BENGAL**

1. **A-de-cuba** : Darjeeling division of W.B. Medium maturity. Orange flint kernels. Resistant to foliar diseases, ear rot and stalk rot.

**(VI) JAMMU & KASHMIR**

1. **$C_2$** : Higher elevation of Srinagar valley. Early maturing variety with yellow flint grains. Low input environments.
2. **$C_2$** : Valley of Srinagar. Medium maturing composite with yellow dent grain. Wide adaptability and reasonable level of tolerance to foliar diseases.

3. **Trikuta** : Lower hill of valley and plains of Jammu. Yellow flint and medium maturing.
4. **Nishat** : Valley areas of Srinagar upto 1809 metres. Medium maturing composite with yellow semi-dent to dent. Reasonable tolerance to foliar diseases.
5. **Man-sar** : Higher elevations of Kashmir and plains of Jammu and Medium maturing composite with yellow semi flint grains. High level of resistance to foliar diseases.

6. **$C_6$** : Grain orange yellow-fllint. Maturity above 100 days.
7. **$C_{15}$** : Grain yellow semi-flint. Maturity above 100 days.

**(VII) MAHARASHTRA**

1. **Manjari** : Orange semi-flint grain. Maturity, 90-100 days.
2. **African tall** : Grain white semi-flint and recommended for fodder across the country.
3. **Hunius.**

**(VIII) SIKKIM 1. NLD.**

**(IX) MADHYA PRADESH**

1. **Chandan safed - 2** : White dent, very early maturity like sathi, photo-insensitive and tight husk.
2. **Chandan- 3** : Medium maturity with yellow grain.
3. **Chandan 1** : Medium maturity, yellow seeded, good adaptability. Comparable in maturity and disease resistance to Vijay.

**(X) HIMACHAL PRADESH**

Parvati : Grain orange yellow flint and maturity 110-115 days.

Recommended for hilly areas and plains.

(XII) GUJARAT

**1. Gujarat Makki, early (GM-1) :** Grain white creamy flint and maturity 80-85 days. Recommended for white maize growing areas.

## I RECOMMENDED PACKAGE OF IMPROVED AGRONOMIC TECHNIQUES

A new strain is not a magic seed. The seed is only a catalyst for improvement. We have to change the whole technique of cultivating and fertilizing to produce better crops. Various improved agro-techniques evolved are described.

### Climatic Requirements

The ideal temperature found for germination is 21° C and that for growth 32° C. At low temperature germination takes more time i.e. at 15.5 - 18.3° C the seedling appear above the ground in 8-10 days and at 10-12.8° C, it takes 18-20 days. Under the mild climate it has been observed to grow throughout the year.

Rainfall varying from 50-75 cm during the crop season is conducive to the proper development of maize plant. Thus, maize requires considerable moisture and warmth from the beginning of planting to the end of flowering. In mild climate it can be grown throughout the year. Hot days and cold nights are preferred for best growth.

### Selection of Land

Maize performs well on well drained loams and silt loams containing abundant organic matter and plant nutrients. Maize has been grown successfully from sea level to 2500 m provided it is not subjected to frost. Maize is grown in all kinds of soils ranging from light sandy loam to deep clay provided pH is between 6 and 8.

### Land Shaping and Seed Bed Preparation

For maize, preparation of fine seedbed is essential. Deep ploughing is beneficial to maize. Moolani and Hukkeri (1965) recorded significantly the highest grain and straw yield at 24-26 cm depth of ploughing with tractor drawn MB plough over 6-8 cm depth of ploughing with country plough and 12-14 cm depth of ploughing with MB UP-2 plough (bullock drawn). The 24-26 cm depth of ploughing significantly reduced quantity of weeds at harvest when compared to 6-8 cm and 12-14 cm depth of ploughing (Table 1).

Table 1 : Effect of depth of ploughing on the grain and stover yield of maize, S:G ratio and quantity of weeds at harvest.

| Depth of ploughing | Grain (q/ha) | Stover (q/ha) | S : G | Weeds (q/ha) |
|---|---|---|---|---|
| 6–8 cm (country plough) – cross wise | 10.45 | 25.27 | 2.42 | 15.6 |
| 12–13 cm MB UP (bullock) | 12.65 | 27.52 | 2.17 | 15.2 |
| 24–13 cm (Tractor MB) | 17.64 | 41.98 | 2.38 | 11.8 |
| S EM ± | 1.24 | 2.25 | – | 0.93 |
| CD at 5% | 3.73 | 6.74 | – | 2.80 |

Source : Hoolani, MK & Hukkeri, SB (1965). Indian J. Agron. 10(1) : 45-47.

The field has to be brought to a fine tilth to a depth of 15-20 cm and must contain adequate moisture for germination and growth. Many soils in developing countries have a very tight compact sub soil that has developed because of shallow cultivation. Use of a subsoiler once every 3-4 years accomplishes the desired effect and breaks this up. Tillage operations following ploughing should consist of discing. In very heavy soils discing more than once may be needed. Each discing should be followed by planking to get a seedbed free of clods and weeds. Tillage should be performed at a moisture level sufficient to prevent complete drying out. While preparing the field for maize, levelling must not be overlooked as maize is very sensitive to water-logging. Therefore, micro- levelling of field should be undertaken by the farmers to ensure better water management and to reduce water logging condition.

## Plant Population and Its Pattern

The importance of adequate plant population to get the maximum yield need not be over emphasised. Experimental data (Table

2) obtained have indicated that 60-65 thousand plants/ha at harvest are near optimum. To achieve this, seed must be dibbled at a distance of 18-20 cm in rows 75 cm apart (Table 3). About ten to twenty percent higher than the desired plant population is normally required to compensate for various field losses between planting and harvesting. About 20-24 kg/ha seed will suffice depending upon the seed size. It has been observed that there is a drastic reduction in the proportion of barren plants with the increase in spacing.

Table 2 : Influence of plant population on grain yield of maize (Average in each zone represent 2–4 locations in each year)

| Plant population (000/ha) | Zone I | Zone II | Zone III | Zone IV | Average (q/ha) |
|---|---|---|---|---|---|
| 25 | 39.3 | 25.8 | 25.6 | 28.8 | 29.9 |
| 50 | 50.7 | 32.7 | 31.1 | 31.5 | 37.5 |
| 75 | 52.5 | 34.4 | 34.0 | 35.8 | 39.2 |

Source : Anonymous (1965–74) Annual Report AICMIP, ICAR.

Table 3 : Responses of hybrid and local maize to plant population and row spacing in rainfed areas (q/ha)

| Row spacing (cm) | Hyderabad | Dharvar |
|---|---|---|
| 60 | 32.6 | 37.9 |
| 75 | 34.5 | 36.1 |
| 90 | 27.4 | 35.4 |
| SEM ± | 0.9 | 1.2 |
| CD 5% | 2.8 | NS |

Source : Anonymous (1974) Annual Report, AICMIP, ICAR.

Under rainfed conditions a plant population of 50 thousand/ha appears to be near optimum. Shukla *et al.* (1978) reported that grain yield of maize significantly reduced by 6.9 and 9.5 q/ha by increasing plant population from 50 thousand/ha to 60 and 70 thousand/ha, respectively. Of course, there was slight increase in straw yield as density was increased (Table 4). One should remember that maize crop does not tiller to compensate for the lower stand.

Table 4 : Effect of plant population under rainfed condition on grain yield of maize (Vijay)

| Population | Grain yield (q/ha) | | | Stover (q/ha) | | |
|---|---|---|---|---|---|---|
| 000/ha | 1974 | 1975 | Average | 1974 | 1975 | Average |
| 50 | 54.8 | 30.2 | 42.5 | 96.3 | 76.5 | 86.4 |
| 60 | 46.6 | 34.7 | 35.6 | 97.4 | 68.9 | 83.2 |
| 70 | 43.0 | 23.0 | 33.0 | 98.0 | 78.9 | 88.5 |
| CD 5% | 5.7 | 1.9 | – | NS | NS | |

Source : Shukla *et al.*, (1978). Indian J. Agron. 23 : 187.

Seed size in relation to yield: Seed size of less than 7.1 mm gave high yields than higher size (Table 5). Seed treatment with JAL-SHAKTI (5%) has given more yield in North Bihar only.

Table 5 : Effect of seed size on the grain yield of vijay maize

| Treatment | Grain yield (q/ha) | |
|---|---|---|
| | 1975 | 1976 |
| Bulk size | 46.2 | 34.9 |
| Large ( > 7.9 mm) | 42.2 | 32.4 |
| Medium ( 7.1–7.9 mm) | 43.6 | 33.7 |
| Small ( < 7.1 mm) | 50.0 | 38.5 |
| CD 5% | 4.8 | 1.7 |

Source : Dhillon *et al.* (1980) Indian J. Agron 25 : 244–249.

METHOD OF SOWING : Plant across the slope. Shallow planting on heavy, damp, cool and crusted soils have proved better. Deep planting should be practiced on dry and warm soils. In warmer situations the seed may be planted as deep as 20 cm so that kernel may have sufficient moisture for germination in spite of the effect of stronger sunshine. Planting done by one of the following method gives the best results :

(1) Planting on the side of a ridge in high rainfall situation and on land which is not uniform.

(2) Plant in a flat bed with no earthing up.

(3) Planting in the furrow in low rainfall areas

(4) Planting on the flat bed and earth up after 40-50 days planting.

Seed is either drilled or dibbled. Ridges are also useful for irrigation and drainage purposes.

## Time of Planting

Middle of June to end of July in general. Adjust the date of planting in such a way as to avoid moisture stress at the crirical stage of flowering, particularly under rainfed conditions for post-monsoon sowing.

Table 6 : Influence of date of planting on grain yield of maize (q/ha)

| Date | Kalimpong | Dholi | Chind wara | Hyder abad | Godhra |
|---|---|---|---|---|---|
| D1 | 64.0 (–)** | 42.5 (O) | 59.2 (–) | 46.0 (–0) | 38.1 (–) |
| D2 | 53.2 (108) | 43.1 (–) | 59.5 (–) | 36.4 (87) | 37.5 (6) |
| D3 | 40.2 (119) | 40.3 (106) | 48.5 (110) | 38.3 (35) | 26.1 (60) |
| D4 | 39.2 (83) | 29.5 (95) | 43.2 (82) | 36.9 (29) | 27.6 (35) |
| D5 | 24.9 (98) | 26.6(78) | 37.8 (73) | 30.0 (15) | 22.6 (39) |

| | D1 | D2 | D3 | D4 | D5 |
|---|---|---|---|---|---|
| Kalimpong | 10/4 | 20/4 | 30/4 | 10/5 | 20/5 |
| Dholi | 25/5 | 05/6 | 15/6 | 25/6 | 05/7 |
| Chindwara | 3–8/5 | 13–16/6 | 23–28/6 | 3–8/7 | 13–18/7 |
| Hyderabd | 20/5 | 01/6 | 12/6 | 21/6 | 01/7 |
| Godhara | 09/6 | 19/6 | 29/6 | 09/7 | 19/7 |

**( ) Kg grain reduction/day

Source : Anomjmous (1964–80) Annual Report, AICMIP, ICAR

By changing the date of planting it is possible to avoid the problems of diseases or insects. The maize should be planted around 10th of April at Kalimpong (W.B.), 20th May at Bajaura (HP), 25th May at Hyderabad (AP), 10th of June at Chindwara (MP), Godhra (Guj.) and Pantnagar (UP) and around 15-20th June at Delhi, Ludhaina (Punjab) and Rajasthan (Table 6). Thus at all locations these dates are about 10-20 days earlier than the normal onset of the monsoon, which gives an opportunity for better management. Sowing done 10-15 days earlier than the usual postmonsoon sowing, with irrigation facilities, have given upto 60% increase in yield. The reason for higher yields obtained from early planting in developing tropical countries is that early planting helps to avoid periods of excess rainfall during early stages of growth when the plants are highly sensitive to excess soil moisture. At some places early planting also helps to avoid moisture stress at critical stage of flowering.

For the early summer crop, the planting should be completed as early as possible otherwise the crop would come into flowering during hot months of March and April. Higher atmospheric temperature cause dessication of pollen, resulting in poor seed set. Moreover, under high temperature, the endosperm ripens quickly thus reducing size of the grain and hence the yield.

## Weed Control and Inter Tillage

Proper and timely control of weeds is one of the most important factors determining success, particularly of *Kharif* maize cultivation. The following practices should be followed which may gradually minimise incidence of weeds :

(1) Use weed-seeds free seed

(2) Use well decomposed FYM

(3) Remove weeds from the cropped area by all means before seeds are shed and take precaution in not allowing weeds come through water, FYM and machinery, etc.

Remember the saying 'One year of seeding of weeds is 7 years of weeding .

(a) Critical stages of crop weed competition : Timely control of weeds is very important particularly of summer maize cultivation. Moolani (1965) reported significant reduction in grain yield of maize as compared to weed free check when the weed *Amaranthus viridus* was removed 10 weeks after planting and onward. The weed removal 10 weeks after planting reduced grain yield to 84.3 per cent of weed free check. There was significant increase in weed dry matter with the advancement of time of weed removal after planting (Table 7).

Table 7 : Grain yield of maize as influenced by the time of weed removal-*Amaranthus viridis*

| Time of weed removal after planting | 1959 | | 1960 | | Mean grain | Grain yield as % of check |
|---|---|---|---|---|---|---|
| | Grain | Weeds | Grain | Weeds | | |
| | (q/ha) | | | | | |
| Weed free | 68.7 | 0.0 | 70.4 | 0.0 | 69.5 | 100.0 |
| 4 weeks | 67.2 | 2.8 | 66.1 | 2.3 | 66.6 | 96.5 |
| 7 weeks | 66.3 | 3.3 | 65.5 | 4.1 | 65.9 | 95.4 |
| 10 weeks | 59.2 | 9.6 | 57.2 | 11.9 | 58.2 | 84.3 |
| 13 weeks | 56.2 | 12.6 | 53.6 | 15.5 | 54.9 | 79.5 |
| 16 weeks | 51.7 | 16.3 | 49.5 | 19.4 | 50.6 | 73.3 |
| Crop maturity | 50.6 | 18.6 | 47.4 | 22.5 | 49.0 | 10.9 |
| SEm ± | 1.99 | 0.29 | 3.46 | 0.57 | 2.71 | – |
| CD 5% | 5.94 | 0.83 | 10.16 | 1.69 | 7.97 | – |

Source : Moolani, MK (1965). Indian J. Agran. 10 (3) : 313-318

It has been observed that weed crop competition in maize is most critical during the first 40-45 days of growth (Gupta and Lamba, 1978). Sandhu and Gill (1973) also observed that delaying weed removal beyond two weeks caused appreciable reduction in growth as well as yield of the crop. They were of the opinion that maximum competition occured during the period of 2 to 6 weeks after sowing which resulted in almost 50% of the total loss caused by weed infestation during the whole crop period.

(b) Chemical weed control : Atrazine and simazine herbicides are highly effective, particularly for the control of broad-leaved and annual grassy weeds within the row and near the base of the plants, and for types of lands and seasons where the cultural operations cannot be done in time. However, atrazine is prefered because of higher solubility, better results in drier regions and its suitability as a post-emergence spray. It seems to have less carry over possiblities for the next crop and is more effective on certain weeds. Application of 0.5 kg (a.i.)/ha of atrazine to maize with two inter cultivations has been found to be satisfactory (Table 8). Weed control is excellent upto 4-5 weeks.

Use of herbicides has icreased the yield by 10-45% in various parts of the developing tropical countries. For inter-crop with pulse alachlor@1loRL/ha or pendamethyl oxidizon @ 0.50 - 0.75 kg/ha and in rainfed area oxadizon @0.50kg/ha were found to be effective. Thakur *et al.* (1990) reported that pre-emergence application of pendimethalin @ 1.5 kg a.i./ha resulted in comparatively higher maize yield (68.4 q/ha) and uptake of N P K followed by hand-weeding twice (60.5 q/ha) and pre + post emergence combination of atrazine and glyphosate (57.9 q/ha).

Table 8 : Influence of cultural and chemical weed control on maize

| Treatments | Grain yield (q/ha) | | |
|---|---|---|---|
| | Pan | Hyd | Average |
| Control | 48.9 | 37.3 | 43.1 |
| Hand weeding | 51.9 | 46.5 | 49.2 |
| Atrazine @ 1.0 kg/ha | 56.5 | 47.4 | 51.9 |
| Atrazine @ 1.5 kg/ha | 56.0 | 47.7 | 51.8 |
| Lasso 2L/ha | 56.3 | 40.4 | 48.4 |
| Lasso 4L/ha | 52.8 | 43.6 | 48.2 |
| Linuron 0.5 kg/ha | 55.9 | – | 55.9 |
| Linuron 1.0 kg/ha | 60.4 | – | 60.4 |
| Tafazine 1.0 kg/ha | – | 45.2 | 45.2 |
| Tafazine 1.5 kg/ha | – | 46.8 | 46.8 |
| SEm ± | 2.3 | – | – |
| L.S.D. at 5% | 6.6 | – | – |

Source : Anonymous (1973–75) Annual Report, AICMIP, ICAR.

A higher amount of organic matter and clay in the soil reduces the effectiveness of herbicides but higher moisture and warm temperature increase the efficiency of the sprays. Atrazine can be sprayed even after the maize crop has come up provided that the weeds are not taller than 4 cm. Simazine is not effective after the weeds have established.

The inter-row cultivation when carried out before the weeds are more than 5 cm tall can control the weeds very effectively in between the rows of maize. However, the weeds within the row and near the base of the plants cannot be removed with the implement used for inter-cultivation.

A liquid side dressing of N with 2,4-D for late weed control is becoming popular in developing countries. The N solution not only replaces the usual side dressing and cultivation but kills young weeds on contact. The added 2,4-D holds other weed back. The base of the maize stalk is tolerent to N sprays but the maize leaves are burned by them. This method gives a late weed control that is not possible by other methods. N @ 90 kg/ha in solution + 2,4-D (acid) + wetting agent are sprayed together. Spray when maize is 50-70 cm high. Two sprays at 7 days interval are enough.

Earthing operation was found to affect both grain and stover yield significantly over no earthing (Table 8a). The increase in yields were upto 35.6 and 22.0% respectively. Earthing up also resulted in significantly higher plant stand (2.6 t/ha) alongwith reduction in barren plants against no earthing. Flat sowing followed by earthing up operation produced 6.1 q/ha more grain yield as compared to no earthing.

Table 8a : Effect of earthing on maize yield under rainfed condition Udaipur.

| Treatment | Grain yield (q/ha) | Plant stand (000/ha) | No. of ears (000/ha) | Plant height (cm) |
|---|---|---|---|---|
| Earthing | 12.7 | 65 | 35 | 143 |
| No earthing | 9.3 | 62 | 32 | 128 |
| SEm ± | 0.6 | 0.6 | 1.6 | 4.7 |
| CD 5% | 1.9 | 2.0 | NS | 14.0 |

Jain G.L. and Jangir, R.P. (1976) AICMIP, Agronomy Report, Sukhadia University, Udaipur.

## Extent of Minimum Tillage

There are several ways in which minimum tillage can prove useful. (i) It has been found to be particularly beneficial during dry years when the moisture conserving surface residues retard water runoff and reduce evaporation. (ii) Under this system, the soil structure improves and helps to resist both wind and water erosion. (iii) Hilly, high rainfall regions of developing countries offer ample scope for no- tillage maize cultivation without accelerating erosion. (iv) Regions in developing countries where cultivation of two to three crops of maize per year is being encouraged face a serious problem in respect of shortage of time for conventional tillage between two consecutive crops.

## Manuring and Fertilization

### (a) Application of FYM

The time of application of FYM is important. The partially rotten FYM should generally be applied to the soil from three to four weeks prior to the sowing of the crop. Given sufficient moisture in the soil, this will allow enough time for its decomposition and improve soil structure. If the manure is well rotten, it is advisable to apply it just prior to the sowing. After the manure is carted to the field, it should be evenly spread and worked into the soil soon to avoid loss of introgen. It is desirable to apply 40-50 cartload of FYM or compost per hectare. Gaur (1991) obtained maize grain yield (26.3 q/ha) with the application of FYM 30 days before sowing compared to early or late application timings. Although there is no substitute for FYM for raising good maize crop keeping in view the fertility status of the soil, but this being scarce, it is advisable to use inorganic fertilizers, keeping in view the fertility status of the soil.

### (b) Application of Fertilizer

The fertilizer recommendations are made keeping in view that maize roots take up 40-60% of the available elements only and the major part of the rest will be lost in drainage water or volatilization and may be fixed in soils. Maize is a heavy feeder and consequently exhausts the soil rapidly where it is grown. The larger the yields the greater is the drain of the nutrients in the soil. It has been estimated that a crop producing 50 q/ha grain removes 164 kg N, 33.5 kg $P_2O_5$ and 128 kg $K_2O$ from the soil (Table 8b).

**(1) Nitrogen** : Soils in India, where maize is grown, are unusually very low in organic matter and available N. Hence most of the N requirement of the maize crop has to be met out from the manures and fertilizers. About 15-37 kg grain/kg N are obtained from hybrids for the first 45 kg of N/ha (Table 9). Thus, hybrids respond profitably to about 50% higher level of N and bring 50-100% higher profits from the fertiliation under rainfed conditions. Whereas under irrigated conditions, maize responded significantly to the application of nitrogen upto 120 kg N/ha in all the four zone (Table 10).

Table 8b : Nutrient accumulation in maize at different growth stages.

| Age (days) | Nitrogen | | Phosphorus | | Potassium | |
|---|---|---|---|---|---|---|
| | kg/ha | % | Kg/ha | % | kg/ha | % |
| 0–30 | 4 | 3 | 0.3 | 1 | 6 | 4 |
| 31–60 | 62 | 38 | 9.5 | 27 | 84 | 66 |
| 61–90 | 76 | 47 | 15.3 | 46 | 38 | 30 |
| 91–100 | 22 | 12 | 8.6 | 26 | 0 | 0 |
| Total | 164 | 100 | 33.8 | 100 | 128 | 100 |

Source : Jain, G.L. (1981) Indian Fmg. 31(8) : 9–11

Table 9 : Response of maize to nitrogen under rainfed conditions (Average : 2–3 locations in each year).

| N | Zone I | Zone II | Zone III | Zone IV |
|---|---|---|---|---|
| (kg/ha) | Yield (q/ha) | | | |
| 0 | 25.3 | 16.7 | 16.1 | 11.0 |
| 45 | 39.1 (30.7) | 24.6 (17.6) | 23.2 (15.8) | 19.3 (18.4) |
| 90 | 49.1 (26.4) | 28.8 (12.4) | 28.8 (14.1) | 26.1 (16.8) |

( ) Return of maize grain/kg of N (Absolute increase over control). Source Annual Report, AICMIP, ICAR.

Table 10 : Response of maize to nitrogen with irrigation (Average 2–3 locations in each year).

| N | Zone I | Zone II | Zone III | Zone IV |
|---|---|---|---|---|
| (kg/ha) | Yield (q/ha) | | | |
| 0 | 23.2 | 16.2 | 19.9 | 22.4 |
| 60 | 44.9 (36.2) | 26.4 (17.0) | 28.5 (14.3) | 30.8 (14.0) |
| 120 | 60.9 (31.4) | 32.5 (13.6) | 38.4 (15.4) | 36.9 (12.1) |
| 180 | 71.0 (26.6) | 36.6 (11.3) | 39.5 (10.9) | 39.0 (9.2) |

( ) Return of maize kg grain/kg of N (Absolute Increase over control). Source : Annual Report, AICMIP, ICAR.

Futher, the improved cultivars viz. Hybrids and composites gave better reponse to the application of nitrogen than the local cultivars (Table 11). At 120 kg N/ha level, hybrids, composites & local cultivars gave 18.1, 17.7 and 12.8 per cent higher yield over the respective control plots.

Table 11 : Response of maize (q/ha) to nitrogen with irrigation (Average All India Zones)

| Germplasms | N (kg/ha) | | | |
|---|---|---|---|---|
| | 0 | 60 | 120 | 180 |
| Hybrid | 22.7 | 34.4 (19.5) | 44.3 (18.1) | 48.4 (14.3) |
| Composite | 18.9 | 32.0 (21.8) | 40.1 (17.1) | 47.1 (15.7) |
| Local | 18.6 | 27.3 (14.5) | 33.9 (12.8) | 35.8 (9.6) |

( ) Return of maize grain/kg of N (Absolute increase over control). Source : Annual Report, AICMIP (ICAR).

Economic optimum levels of N are between 125-175 kg/ha at Almora (UP), Bajaura (HP), Delhi, Dholi(Bihar), Hyderabad (A.P.) and Udaipur (Rajasthan). Economic responses upto 240 kg/ha of N have been obtained under favourable conditions at other locations like Solan (H.P.), Srinagar (J & K), Pantnagar (U.P.), Chindwara(M.P.), Arbhavi (Maharashtra) and Godhra (Gujarat). The minimum economic rate has been found to be 100 kg N/ha at 6 as a cost: price ratio which may produce 12-25 kg grain/kg N, while maize hybrids in India on an average yield 12-22 kg grain/kg N and local cultivars gave 9-16 kg grain. Usually 5 kg grain/kg N pays the cost of N (Jain, 1981).

**(2) Phosphorus** : We commonly see early growth response to P even on soils high in available P. Lack of P stunts growth. In early stages of growth a large share of P in the plants comes from the fertilizer. Later on most of it comes from the soil. One possible reason for this behaviour is that unphosphated plants may be stunted only in the early stages while the root system is small. As the extensive fibrous root system develops, the plants may use the P already in the soil effectively.

Very high response to $P_2O_5$ (upto 60 kg/ha) has been observed on cultivator's fields at Chindwara (M.P.), Solan (H.P.), Srinagar

(J&K) and Udaipur (Rajasthan) (Table 12). The response at 30 kg/ha was upto 17 kg grain/kg $P_2O_5$ applied (Table 13). In general, if the soil tests indicate low, medium or high availability of $P_2O_5$, the recommended rates of $P_2O_5$ application are 80,50 and 0 kg/ha, respectively.

Table 12 : Response of maize to phosphorus application on cultivators field

| Treatments | Zone I | Zone II | Zone III | Av yield |
|---|---|---|---|---|
| $P_2O_5$ (Kg/ha) | | (q/ha) | | |
| 0 | 60.3 | 12.7 | 17.3 | 30.1 |
| 30 | 65.8 | 14.2 | 27.6 | 35.9 |
| 60 | 68.3 | 17.8 | 32.8 | 39.6 |
| 90 | 75.9 | 17.1 | 40.1 | 44.4 |

Source : Annual Report, AICMIP, ICAR.

Table 13 : Response of maize to phosphorus under rainfed condition (Udaipur, Kharif).

| $P_2O_5$ (kg/ha) | 1972 | 1974 | Av yield (q/ha) |
|---|---|---|---|
| 0 | 18.2 | 7.2 | 12.7 |
| 15 | 18.4 | 10.0 | 14.2 |
| 30 | 23.8 | 11.7 | 17.8 |
| 45 | 21.8 | 12.4 | 17.1 |
| SEm ± | 1.2 | 0.4 | – |
| CD at 5% | 3.4 | 1.0 | – |

Source : Annual Report, AICMIP, ICAR.

**(3) Potassium :** Response to potash in Kulu valley (HP) was high, the average being 5-6 q/ha which was significant (Table 14). Besides, application of potassic fertilizer (30 kg $K_2O$/ha) resulted in an additional yield of about 2.0 q/ha under late sown conditions. This might be on account of late sowing under rainfed conditions by which time most of the rains in the season were over and sufficient $K_2O$ could not be liberated on account of poor moisture supply. It may be ascribed to the fact that when crop is sown late the rain water is insufficient to fulfil the requirement of K, through soil water. Under such circumstances, if potassic fertilizers are applied they might be helpful in boosting the production of maize.

Table 14 : Response of maize to potash application on cultivators fields.

| Treatment $K_2O$ (kg/ha) | Srignagar (I) | Bajaura (I) | Hyderabad (IV) | Godhara (V) | Av yield (q/ha) |
|---|---|---|---|---|---|
| 0 | 58.0 | 59.3 | 23.7 | 33.8 | 43.9 |
| 50 | – | 64.6 | 25.6 | – | 45.1 |
| 100 | 60.0 | – | – | 37.7 | 48.9 |
| SEm ± | 0.5 | – | – | 0.8 | – |
| CD 5% | 1.3 | – | – | 2.3 | – |

Source : Annual Report, AICMIP, ICAR.

**(4) Time of application :** Under tropics and semi-arid tropics the plant should make the optimum use of the applied elements if the following schedule for their application is followed : (1) quarter to one-third N, all $P_2O_5$ and $K_2O$ should be applied as basal dose, i.e. before planting, (2) half of the remaining N should be applied when the crop is knee high i.e. about 35-40 days after planting and (3) the remaining N shall be applied just before silking and after tasseling (Table 15).

Table 15 : Effect of time of application of recommended nitrogen on maize yield at Ludhiana

| Time of nitrogen application | | | | | |
|---|---|---|---|---|---|
| Planting | 20 days | 30 days | 40 days | Tassel | Yield (q/ha) |
| Full | – | – | – | – | 12.8 |
| 1/3 | – | 1/3 | – | 1/3 | 22.7 |
| 1/5 | – | 4/5 | – | – | 20.8 |
| 1/3 | – | 2/3 | – | – | 20.6 |
| 1/4 | – | 1/4 | 1/4 | 1/4 | 23.4 |
| 1/5 | 1/5 | 1/5 | 1/5 | 1/5 | 23.8 |
| 1/4 | 1/4 | 1/4 | – | 1/4 | 23.2 |
| 1/3 | – | 1/3 | 1/3 | – | 24.0 |
| 1/5 | – | – | 4/5 | – | 17.9 |
| 1/3 | – | – | 2/3 | – | 17.2 |

Source Annual report, AICMIP, ICAR.

Under rainfed condition the amount of fertilization would vary with soil fertility and the amount and distribution of rains. Usually 50-57 kg of N and 25-35 kg of $P_2O_5$ /ha would prove adequate.

(v) Techniques and sources of fertilizer application : The basal dose may be placed in a band 5 cm to the side and below the seed (Table 16). Second and third dose should be top dressed. After top dressing either rake the fertilizer or earth up the soil so that it is incorporated well into the soil. After each application the field must be irrigated if there is no rain.

Table 16 : Effect of method of fertilizer application on maize yield (q/ha) in rainfed area

| Treatment | Delhi | Dholi | Chindwara |
|---|---|---|---|
| Band | 37.4 | 32.7 | 72.1 |
| Broadcast | 32.4 | 30.6 | 67.6 |

Source : Annual Report : AICMIP, ICAR

Deep placement of N was most effective on fine textured, poorly drained soils. On the other hand, on coarse textured, well drained soils depth of placement had no effect on maize yield. Deep placement would also be more effective in dry years.

Ammonium sulphate in the rainy season and urea in the winter season are better fertilizers. Fertilizers like ammonium sulphate nitrate in the standing crop has proved better than ammonium sulphate or urea. A few reports indicate that fertilizer containing $NH_4$ -N might be better than those containing $NO_3$ -N under the condition of heavy rainfall due to increased leaching losses.

Studies on nitrification retardants conducted at Delhi showed (Table 17) that application of neem cake coated urea at sowing increased grain yield by 3.0 and 8.8 q/ha over uncoated and sulphur coated urea, respectively. While, at Udaipur conditions, maize did not respond significantly to the application of nitrification retardents. However, Mehta (1991) observed reponse of maize to the application of neem extract coated urea. He observed that application of NECU improved the plant growth characters, yield attributes and yield of maize over plain urea. It increased the net return by Rs.140/- per hectare over plain urea application (Rs.9451/-).

Table 17: Study on N utilization efficiency through application of nitrification retardents and their application

| Treatment | Grain yield (q/ha) | | |
|---|---|---|---|
| | Delhi | Udaipur | Average |
| All U(sow) | 56.8 | 34.8 | 45.8 |
| 1/4 U (Sow) + 1/2U (30) +1/4 (Tass) | 44.3 | 36.7 | 40.5 |
| SCU (Sow) | 51.7 | 40.6 | 46.2 |
| U+NC (Sow) | 59.8 | 41.9 | 50.9 |
| 1/2 SCU (Sow) + 1/2 U (Tass) | 53.6 | 41.3 | 47.5 |
| 1/2 U (Sow) + 1/2 U (Tass) | – | – | – |
| 1/2 U + NC (Sow) + 1/2 U (Tass) | 52.5 | 43.2 | 47.9 |
| 1/4 Comp (Sow) + 1/2 U (30) + 1/4 U (Tass) | 43.4 | 35.4 | 39.4 |
| All U (30) | 43.3 | 35.4 | 39.4 |
| SEm ± | 2.6 | 2.4 | – |
| CD at 5% | 7.3 | NS | – |

U = Urea, SCU = Sulphur coated urea, Comp = Compost, NC = Neem cake; Sow = At sowing, 30 = 30 days after sowing, Tass = at tasselling.

Regarding $P_2O_5$, complex fertilizers (DAP, IFFCO mixture, etc.) containing N and at least 50% water soluble P may have an advantage over single superphosphate under high pH soil condition. Urea ammonium phosphate and ammonium nitrate phosphate offer better handling properties. Complex grade fertilizer like above can be mixed with seed before drilling without any harmful effect on seed germination.

Amongst the various micronutrients investigated mostly on sandy and alkaline calcareous soils and freshly levelled fields, zinc deficiency has been observed in some parts of India- J & K, Haryana, Punjab, Delhi, UP (Tarai), Gujarat and AP. Response to zinc applied through soil or foliar sprays, has been obtained upto 25 kg/ha zinc sulphate (Table 18). The response was upto 20 q/ha grain. It also enhanced silking upto 10 days at Ludhiana (Punjab).

Table 18 : Response of maize (q/ha) to levels of zinc

| Zinc sulphate | Zone I Srinagar | Zone II Ludhiana | Zone III Dholi (R) | Zone III Godhra | Average |
|---|---|---|---|---|---|
| 0 | 32.7 | 18.3 | 33.4 | 20.5 | 28.9 |
| 10 | 36.5 (38) | 46.6 (283) | 33.6 (2) | 27.6 (71) | 32.6 (37) |
| 20 | 42.3 (48) | 45.9 (138) | 34.8 (7) | 31.8 (56) | 36.3 (37) |
| 40 | 47.5 (37) | 45.8 (68) | 36.5 (8) | 25.5 (13) | 36.5 (19) |
| SEm ± | 0.8 | 2.3 | 0.7 | 0.7 | – |
| CD at 5% | 2.2 | 6.6 | 2.1 | 2.0 | |

() Return of maize kg grain/kg zinc sulphate.

Source : Annual Report, AICMIP, ICAR

*Bacterial inoculation :* Seed treatment with *Azotobacter* increased grain yield considerably (8-15%) as compared to untreated plots. Similarly, seed inoculation with phosphobacteria increased grain yield over uninoculated seed.

## Water Management

Maize is sensitive to excess water as well as moisture stress. It is desirable to plant it on ridges in the field after its establishment. It is convenient and economical to provide irrigation water to maize in the furrows which may serve as drainage channel, if and when required. The optimum depletion level of soil moisture for maize is around 50% of availability from the surface 30 to 60 cm depth on sandy loam to clay loam soils. Generally, in lighter soils frequent and in heavy soils infrequent and heavy irrigations are recommended for best results. Withholding of irrigation is recommended in case of about 20-25 cm of rainfall in a week.

Proper scheduling of irrigation should also have some practical consideration. In areas having ample and cheap water resourses with assured supply throughout the crop season, it may be poosible to schedule irrigation as and when required to meet the full water needs to plants. On the other hand, in areas where water resources are scarce and where high value crops are grown, precise approaches for irrigation are desirable.

Adequate moisture to prevent wilting of maize plants is essential throughout the growing season. It has been shown that moisture stress occuring at various vegetative and reproductive stages of growth and development of maize plant may reduce the grain yield. The extent of yield reduction depends not only on the severity of the moistrue stress but also on the stage of plant development when the stress occurs and to some extent on climatic conditions.

The water requirement of maize, on account of its broad leaves and tall growth, is very high. A good crop of maize requires a rainfall or irrigation of about 1.0 to 1.2 m per hectare during its life cycle, most of this is needed during the growing period. If the leaves roll during the middle of the day, more water is needed. Irrigation can be given from the water harvested from catchment in the field and adjoining areas.

One irrigation should be applied when plants are approximately 6"-8" tall. This will substantially stimulate growth even though the soil is moist. Irrigation also is necessary with each fertilizer application as top dressing if there is no rain. Further, irrigation should be given at silking and tasseling period as crop needs maximum water at this stage.

Experimental data have shown that soil moisture depletion to the wilting point for periods of one or two days at tasselling stage resulted in yield reduction of as much as 22% while such deficits for 6-8 days at this stage reduced yield about 50% or more. However, marked varietal differences do exist to moisture stresses. Based on the available information, maize should be irrigated at 0.5 atmospheric tension (15 cm soil depth).

Sharma and Upadhyaya (1973) recorded the highest grain yield of maize (63.1 q/ha) with four irrigations applied at the 6th leaf, late knee high, tasselling and silking stages of the crop (Table 19).

Naidu and Reddi (1975) observed significant effect of interaction between nitrogen levels and moisture regimes on grain yield of maize. At 100 and 150 kg N/ha levels, depletion of soil moisture beyond 40 per cent significantly reduced grain yield. Whereas at 200 kg N/ha level, significant reduction in grain yield was observed even at the depletion of soil moisture beyond 20 per cent. It may be concluded that at levels of 150 and 250 kg N/ha the crop

Table 19 : Effect of irrigation treatment on grain yield of maize on water use by the crop.

| Treatment | 6th leaf | Early knee high | Late knee high | Tasselling | Silking | No. of irrigation | Grain (q/ha) | Water (mm) | kg grain/ mm water |
|---|---|---|---|---|---|---|---|---|---|
| $T_1$ | + | + | + | + | + | 5 | 49.0 | 551.5 | 8.8 |
| $T_2$ | – | + | + | + | + | 4 | 55.6 | 442.3 | 12.5 |
| $T_3$ | + | – | + | + | + | 4 | 63.1 | 438.0 | 14.4 |
| $T_4$ | + | + | – | + | + | 4 | 50.7 | 455.6 | 11.1 |
| $T_5$ | + | + | + | – | + | 4 | 32.7 | 473.3 | 6.8 |
| $T_6$ | + | + | + | + | – | 4 | 46.8 | 456.4 | 10.2 |
| $T_7$ | + | + | + | – | – | 3 | 31.9 | 410.6 | 7.8 |
| $T_8$ | + | + | – | + | – | 3 | 46.6 | 411.8 | 11.3 |
| $T_9$ | Irrigation at 0.8 amt. SMT | | | | | 8 | 42.3 | 553.4 | 7.6 |
| $T_{10}$ | + | – | – | – | – | 1 | 26.7 | 324.1 | 8.2 |

(+) Irrigation given, (–) Irrigation withheld, CD (5%) 13.1 ** SMT (Soil Moisture Tension was measured with Irro-meters installed at 20 cm soil depth.

Source : Sharma and Upadhaya (1973). Indian J. Agron 18 (2) : 220–221 should be irrigated at 40 and 20 per cent soil moisture depletion, respectively (Table 20).

Table 20 : Grain yield (q/ha) of Deccan Hy as affected by levels of N under different moisture regimes.

| Treatment | Moisture regimes (soil moisture depletion) | | | | |
|---|---|---|---|---|---|
| (N kg/ha) | 20 % | 40 % | 60 % | 80 % | Mean |
| 100 | 42.1 | 41.6 | 30.2 | 21.7 | 34.0 |
| 150 | 49.1 | 46.7 | 26.9 | 23.9 | 36.6 |
| 200 | 51.5 | 46.2 | 24.6 | 19.2 | 35.4 |
| Mean | 47.6 | 44.8 | 27.2 | 21.6 | |
| CD 5% (Moisture regime) : 3.7 | | | | | |

Source : Naidu & Reddi (1975) : Indian J Agron 20 (3) : 293–94

For irrigation, preparation of furrows and ridges proved to be very useful because they serve a dual purpose to irrigation and drainage. Use of sprinkler systems, particularly in *Rabi* and summer, when the water is scarce, proved even better. This is particularly useful where the slope of land is uneven because it prevent soil erosion. Alternate furrow irrigation saved 38% water without appreciable reduction in grain yields.

## Drainage

Making arrangement for drainage might be required in the field where the water table is high and it is not possible to grow a good crop of maize without having artificial drainage. However, in many fields, where the water table is not very high provision of surface drainage improves the crop yield substantially.

Waterlogged fields at flowering have exhibited increased stalk rot damage. Drainage encouranges deeper and more extensive root system in maize plants. Thus, the crop can draw upon a large reservoir of soil moisture to carry it over to the periods of inadequate rainfall which sometimes occur in India during monsoon season. These deep roots will also be useful in supplying water to the crop during the maturation period after the monsoon rains cease.

The surface drainage should be provided by opening channels

at a distance of 20-25 metres across the slope of maize field. These drains should finally be connected with a large drain through which water will ultimately flow out of the field.

## Mulching

Under drought conditions maize yields have been increased markedly by the use of a surface mulch particularly under *barani* cultivation (Table 21). During the dry periods the moisture in the surface 2" of soil under the mulched area was higher and temperature was lower than for the unmulched plots. The soil in the mulched plots dried out slower than in the unmulched plot due to reduced evaporation. Also, the soil in the mulched plot appeared to be soaked more by a moderate rains than the unmulched soil. Foliar spray of 6% kaolin has been found to be effective in AP, HP, Karnataka, Maharashtra and TN.

Table 21 : Effect of mulch application on maize under rainfed conditions.

| Mulch application | Grain yield (q/ha) | | | |
|---|---|---|---|---|
| | Del | Uda | Lud | Av. |
| Mulch | 43.3 | 12.0 | 43.8 | 31.1 |
| No Mulch | 38.9 | 10.6 | 39.2 | 29.6 |
| S Em ± | 1.0 | 0.3 | 2.4 | – |
| CD at 5% | 2.9 | 1.1 | 4.1 | – |

Source : Annual Report, AICMIP, ICAR

## Lodging and Its Prevention

Apart from the choice of sturdier cultivars and reducing the plant densities, the farmer can exert some influence by way of management practices. For instance, he can slow down early growth if the stand density is high. He can do this by applying N slightly later than usual or by increasing the deressing. There are instances known in which the strength of the straw was increased by late application of N.

If lodging occurs early e.g. at about flowering time, a yield loss of 500-1500 kg/ha is not exceptional. The later the lodging occurs the smaller is the yield loss.

## Intercropping

Intercropping is one way of increasing land productivity in terms of biological yield per unit area. It has become an essential component in modern agriculture under the conditions of assured and adequate input supply. It has been found profitable to grow a number of intercrops in between the inter space of maize rows. Plant characteristics of maize are favourable to accomodate intercrops without much competition or yield decline. The following points need attention while intercropping in maize :-

### (a) Choice of Intercrops

Kalra and Gangwar (1980) reported that grain yield of maize intercropped with mung was significantly higher than pure maize. In addition to increase in maize yields a bonus *mung* yield of 5-6 q/ha was obtained (Table 22) Intercropping of maize either with cowpea, *rajmah,* or soybean in 2:1 ratio was most productive in terms of maize equivalent yield and declined thereafter, with increase in the legume proportion, though maintained its superiority over sole planting of maize (Shah *et al.*, 1991).

Table 22 : Effect of intercropping (Maize + Legume) on maize yield

| Treatments | Maize (q/ha) | | Associate crops (q/ha) | |
|---|---|---|---|---|
| | 1973 | 1974 | 1973 | 1974 |
| Maize alone | 15.2 | 12.9 | – | – |
| Maize + Soybean | 15.7 | 13.4 | 5.1 | 2.5 |
| Maize + Blackgram | 15.5 | 15.9 | 4.4 | 8.3 |
| Maize + G. Nut | 17.7 | 14.6 | 10.1 | 11.6 |
| Maize + G. Gram | 19.1 | 16.0 | 3.6 | 5.3 |
| Maize + Cowpea | 19.3 | 13.1 | 3.6 | 6.2 |
| CD 5% | 2.6 | 2.5 | – | – |

Source : Kalra, GK & B. Gangwar,. (1980) Indian J. Agron 25 (2) : 181–85

### (b) Choice of Cultivars

It has been generally observed that inter-crops yielded more in association with shorter cultivars of maize as against taller. On the other hand tall cultivars of maize suffered the least in competition with intercrops. Khera *et al.*, (1979) obtained better seed yield of black gram when they intercropped it with dwarf J-605 cultivar of maize with erect leaves than when intercropped in the tall leafy vijay composite. Similarly, higher intercrop yields were recorded at Chindwara with dwarf cultivar (H.175) of maize as compared with tall cultivar ($H_1$) during the experiments conducted from 1978 to 1981. During trials conducted at Delhi the semi-spreading type M-13 cultivar was identified as the most suitable cultivar of groundnut for intercropping with maize (Anon., 1980, 1981).

## (c) Crop Geometry and Methods of Intercropping

There are conflicting reports about the ideal row spacing for maize based intercropping system. It has not been conclusively established whether single rows or paired rows of maize are more suitable for intercropping system. Jaganathan *et al.* (1979) found 60 cm row spacing of maize to be better than 75 cm row spacing for intercropping soybean. Khera *et al.*, (1979) found uniform spacing of 75 cm to be a better arrangement than either 90 cm row spacing or paired row spacing (30/120 cm or 30/180 cm) arrangement for intercropping of blackgram. Moolani (1965) recorded higher total production in 75 cm uniform row spacing than in 45/105 cm paired row arrangement for maize/soybean intercropping system. Rathore *et al.* (1978) found paired row cultivation of maize with 30 cm distance in between the rows to be a better arrangement than the same number of rows of maize in single row system, with inter-row space of 60 cm for intercropping blackgram with maize under rainfed condition leading to better economic returns. They observed that the beneficial effect by pairing maize rows and intercropping the interspace was probably due to reduced cost of inter-cultural operations, the need of which was obviated due to dense canopy development by legume. There was no significant difference in yield of maize intercrops between 75 or 90 cm row spacing at Delhi, Udaipur and Pantnagar (Anon. 1974). However, at Solan and Dholi, there was slight advantage in maize yield in wide rows with the intercrops. At both these locations maize yield in pure stands at 90 cm row spacing was slightly less than that in 75 cm row spacing. Reddy *et al.* (1980) recorded 8 to 12% higher grain yield in maize in paried row spacing against conventional equidistant sowing.

Similarly maize grown alone at a spacing of 60 cm x 30 cm yielded 29.3 q/ha compared with the total yield with soybean (33.3 q/ha) or with urd (31.4 q/ha) when grown in alternate rows. Thus inter- cropping of maize with legumes such as soybean, rajmahs (*Vicia faba*), blackgram, green gram, groundnut, sesame, cowpea, cluster bean or lablab beans had no adverse effect on grain yields of maize and gave additional yield compared with maize in pure stand. Maize grown with cowpea in alternate rows drilled reduced soil water, N and P losses considerably as compared to maize grown as pure or with cowpea intersown (Anon. 1964-88).

Maize intercropped with soybean (4:1) recorded the highest mean biological maize equivalent compared to its sole planting or different combinations of legumes. Maize + *rajmah* recorded the highest mean LER (1.81) followed by maize + cowpea (1.74) and maize + soybean (1.59) all sown in the ratio of 2:1. Maize + rajmah (2:1) recorded the highest mean monetary advantage. Highest B:C ratio (5.09 and 4.45) was obtained with maize + soybean (4:1) (Shah *et al.*, 1991).

Singh (1968) recommended cross sowing of maize forage with cowpea. He noted an increase in yield of forage by 120 q/ha with this method over sowing of maize and cowpea in alternate lines. The percentage increase was 32. The average yield of the alterante sown maize and cowpea was 375 q/ha as against 495 q of maize and cowpea when cross planted. He attributed improved yield of cross planted crop to less competition between cereal and legume and also comparative freedom from weeds due to the dense stand of legume and non-legume. One important precaution to be taken in this method is to ensure satisfactory and quick germination of non-legume component as this may suffer badly due to heavy inter twining effect of legume as germination and growth of legume is faster than that of non-legume.

Time of sowing of intercrops : Some preliminary work was initiated at Delhi to find the best time of sowing of intercrops (Anon.. 1981). While the time of sowing (a week earlier, alongwith maize or a week later) of black gram did not affect the yield of either black gram or maize. Groundnut sown a week earlier than maize yielded higher by 2-3 q/ha.

### (d) Nitrogen and Phosphorus Management in Intercropping System

Rao (1985) and Rao *et al.* (1987) recommended 80 kg N and 30 kg $P_2O_5$/ha for maize intercropped with mung. The said treatment gave 4.5 and 3.5 q/ha extra maize grain yield over maize alone and maize intercropped with cowpea, respectively. While in terms of maize equivalent grain yield, maize intercropped with cowpea and fertilized 80 kg N + 30 kg $P_2O_5$/ha gave equivalent grain yield of 53.7 q as against 48.3 q and 34.0 q/ha obtained from maize intercropped with mung and maize alone, respectively.

## Plant Growth Regulators

Field investigations in early years of 1980 on maize showed that Atonik and Mixtalol application did not affect grain yield and other characters. In the following year 1981, Mixtalol and Atonik interacted at 30 kg N/ha for height. Besides, they also increased the yield and uptake of N,P and K. Mixtalol, however, showed superiority over Atonik (Jain *et al.*, 1989). Further, investigation showed that application of Mixtalol followed by Miraculan and Atonik increased grain yield significantly over tap water. The respecitve increases were of the order of 25.4, 17.9 and 12.9 percent during 1982 and 10.0, 4.8 and 4.3 percent during 1983. Amongst the growth stimulants, seed treatment and two foliar sprays of Mixtalol (2 ppm) gave the highest grain yield during both the years and the yield was statistically significant over all other treatments during both the years. The average increase in total dry matter due to Mixtalol were of the order of 11.96 and 8.81 percent over tap water and Atonik, respectively Table 23 (Jain *et al.*, 1987).

Table 23 : Response of maize to some new plant growth stimulants (Jain *et al.*, 1987).

| Growth stimulant | Grain (q/ha) | | Total DM (q/ha) | |
|---|---|---|---|---|
| | 1982 | 1983 | 1982 | 1983 |
| Tap water | 43.9 | 28.9 | 111.7 | 67.8 |
| Atonik | 49.6 | 30.1 | 113.0 | 69.7 |
| Mixtalol | 55.0 | 31.8 | 124.9 | 74.0 |
| Miraculan | 51.8 | 30.3 | 119.9 | 70.8 |
| SEm ± | 1.25 | 0.36 | 3.43 | 0.64 |
| CD 5% | 3.62 | 1.05 | 9.89 | 1.84 |

Mehta (1991) noted that application of growth stimulant Vipul as granules and foliar spray significantly improved productivity of crop. They increased the grain, stover and biological yield. The grain yield increase was to the extent of 7.1 and 12.1 percent, respectively over control (37.41 q/ha). There was no significant difference in yield due to application of Vipul as foliar spray over granules. The application of vipul granules and foliar spray increased return to the tune of Rs. 579 and 1042/ha over control.

## Harvesteing and Threshing

The maize crop is ready for harvest even when the stalks and leaves are somewhat green while the sheeth has dried and turned brown.

When hybrid and composite maize reaches its maximum dry weight a black line forms at the base of each kernel. When the kernel stops growing, the placenta tissue cells collapse and turn back, shutting of any additional nutrients from the cob. This layer usually develops at about 30% moisture with normal maturity.

Threshing is done by shellers operated by hand or power. These dehusker-cum shellers which have intensive labour-saving devices need popularization.

## II IMPORTANT PESTS OF THE CROP AND THEIR CONTROL MEASURES

Total number of insect pests attacking maize exceeds sixty. Among them the stalk borer- *Chilo partellus* (Swinhoe) and *Sesamia inferens* (Walker) are most serious. *Chilo* occurs throughout the year in the country, while Sesamia occurs as a major pest of *Rabi* season in Peninsular India. The estimated loss primarily due to *C. partellus* varied from 26.7% to 80.4% in different agro-climatic regions of India.

The loss due to *S. inferens* in *Rabi* season varied from 25.7 to 78.9%. Other pests are of minor importance though aphids, ear worms, weevils, etc. occur as sporadic pests in particular areas only. Most of the plant protection measures adopted for the control of stalk borer will incidently control many of the minor pests also.

Among different control measures, development and use of

the resistant cultivars is the most dependable method. As a result of screening cultivars for resistance, a source of resistance to *C. partellus* was located in Antigua Gr. I. This has been used by maize breeders in Ganga-5 and Jahawar composite for incorporating borer resistance. Another maize hubrid H-207 from colombia has recently been found to show fairly good level of resistance to *C. partellus*.

Among the maize cultivars screened for resistance to *S. inferens*, Antigua Gr. I, EH 4207, Cuba, 11 J, Him 123, EH 4030, Comp. A 5, JEL 12, VL 2 etc. were found to be fairly resistant to *C. partellus*.

### 1. Stem Borer (*C. partellus S.*)

Mechanical measure : Pull out all dead hearts and destroy them in order to kill the lingering stages of the pest in the stubbles. Plants showing severe borer injuries may be removed at thinning.

Cultural : Remove all the grasses particularly wild sorghum as it serves as an alternate host for the borers of maize. Plough up and burn stubble after harvest. Sowing of crop in first week of July evades borer infestation.

Biological : *Trichogramma minutum* : May be used to parasitise the eggs of the borers. *Apanteles* sp. have been recorded to parasitise the larval stages of the pest.

Chemical : Spray endosulfan (35 EC) at 0.1% emulsion, lindane 20EC at 0.05% or carbaryl 50 WP at 0.1% @ 450 1/ha at 15 days after sowing followed by an application of 4% granules put in whorls @ 5.0 to 7.5 kg/ha, 15 days after the first application in plant whorls. Apply carbofuran 3% granules @ 5 kg/ha in 15 days old plant whorls.

### 2. White Grubs

Hand collection of adults and killing them in kerosinised water help in reducing future generation. (ii) Spray carbaryl 50 WP 0.1%; monocrotophos 36 WSC 0.05% or quinalphos 25 EC 0.03% on host trees of betters. (iii) Use phorate 10% or quinalphos 5% granules in rows @ 25 kg/ha before sowing the crop in endemic areas.

### 3. Termites

Mix 5% aldrin @ 25kg/ha in the soil before sowing in termite prone areas.

### 4. Wireworms

Dust 5% aldrin dust @ 25 kg/ha in soil before sowing the crops.

### 5. Thrips

Spray phosphamidon 85 SL 0.03%, malathion 50 EC 0.05% or endosulphan 35EC 0.1% at the appearance of foliar infestation.

### 6. Pyrilla

Spray metasystox 25EC 0.03%, monocrotophos 36 WCS 0.03%, carbaryl 50 WP 0.1% or endosulphan 35 EC 0.1% at the time of appearance of the pest.

### 7. Pink Borer (*Sesamia inferens*)

The cultural, mechanical and insecticidal control measures given above for *C. partellus* may also be used for control of *S. inferens*.

### 8. Leaf Aphids

**(i) Mechanical :**

In the early stages when aphids are noted in colonies on a few plants only the affected plants should be removed.

**(ii) Chemical :**

When the infestation is found to be marked insecticidal control has to be adopted. Spray 0.05% endosulphan (1.5 ml/l of water) and repeat it 15 days after if necessary.

**(iii) Biological :**

In nature they are usually controlled by certain lady bird beetles like *Menochilus sexmaculata*. Both the grubs and adults of the beetles are predatory on aphids. They can be put into service by release wherever possible.

### 9. Ear Worms (*Heliothis armigera*)

**(i) Chemical :**

Apply methyl parathion 2% dust @ 25 kg/ha. Avoid growing alternate host crops near the vicinity. Spray methyldematon 25 EC, monocrotophos 36 WSC, phosphamidan 85 SL or endosulphan 35 EC 0.01% at the appearance of pests.

**(ii) Mechanical :**

Chopping of silks after pollination, i.e., after the silk start drying, helps to reduce the infestation as the eggs of this pest are usually laid on stalks.

**(iii) Chemical :**

Dust 5% aldrin @ 25 kg/ha, if necessary the neighbouring fallow area may be dusted with 10% so as to kill the emerging hoppers.

### 10. Surface Grasshoppers (*Chrotogenus* sp.) and grasshoppers (*Hieroglyphus nigroreplites*)

**(i) Cultural :**

As the eggs are usually laid on field bunds and higher elevated grounds trimming of the bunds will help in destroying the eggs of grass hoppers. Avoid growing alternate host crops like Jowar, Bajra, Paddy, etc. in the vicinity.

### 11. Hairy caterpillers, Blister beetle and army worms

Chemical : Dust the crop with methyl parathion (2%) dust @ 25 kg/ha.

### 12. Shoot fly (*Atherigona* sp.)

**Cultural :**

As the fly breeds on rotten material it is advisable to properly cover pits of the rubbish with soil. Use of high seed rate and subsequent thinning of affected young seedlings is also recommended.

**Mechanical :**

Affected seedlings in early stages showing dead-hearts may be destroyed.

**Chemical :**

Apply 10% phorate granules in the furrows before sowing at the rate of 15-20 kg/ha.

### 13. Cutworm (*Agrotis ypsilon*)

They cut the seedlings of maize at ground level particularly in the winter season resulting in wide gaps.

**Cultural :**

Plough the field after harvesting the crop to expose the hibernating pupae to the surface. Intercultivation between rows also helps to expose the larvae and pupae to the surface so as to become a prey to the insectivorous birds.

**Chemical :**

Use 5% aldrin or heptachlor dust @ 25 kg/ha in soil treatment or endosulphan 4% dust @ 25 kg/ha as foliar treatment of seedlings.

## III MAJOR DISEASES OF MAIZE AND THEIR CONTROL MEASURES

It is a general practice to use chemicals for the control of diseases. Other methods often used are the manipulation of cultural practices such as shifting sowing date, altering the frequency or the number of irrigations, etc. In addition to these genetic resistance is used to minimise the losses due to diseases. In case of maize the last mentioned method, owing to economic considerations and efficacy, is more popular. In other crops, the high yielding cultivars have often been observed to have a lower level of tolerance to diseases. However, maize hybrids, synthetics and composites developed in India have a high level of disease tolerance. Utilization of genetic resistance is, therefore, likely to be an effective strategy for the control of maize diseases. According to Ullstrup (1966) "The incorporation of genetic resistance into agronomically desirable hybrids by breeding methods seems to be the most efficient and permanent means of controlling maize diseases". As a general precaution copper based fungicides should be avoided as maize is copper shy.

**1. Rust (*Puccinia sorghi*) :**

Dithane M-45 2.00 kg/ha and Dithane Z-78 2.0 kg/ha in 1000 litres to be effective. Spray as soon as first symptoms of the disease appear or at knee high stage of the crop growth.

**2. Brown stripe downy mildew (*Sclerophthera rayssiae* var. zeae):**

The disease can be minimized by spraying mancozeb (0.3%). First spray a month after planting or as soon as the disease symptoms are seen. Other 3-4 sprays should be done at 10-15 days intervals depending on severity of the disease.

Cultural practices : Brown stripe downy mildew (*S. rayssiae* var. zeae) and bacterial stalk rot : Burn crop residue after the harvest. Practice 3 year crop rotation to minimise the initial inocul µm load in the field. Sowing, starting from mid June to July showed that the incidence of the disease decrease with delayed sowing. Thus, sowings done in the second fortnight of June will, therefore, be effective in checking both brown stripe downy mildew and bacterial stalk rot. Incidence of disease becomes less if planting is done during dry season. Select well drained fields for maize planting. Use resistant maize cultivars such as G-5, G- 9, Kisan, Tarun, Naveen, Sweta, Kanchan, etc.

**3. Turcicum and Maydis Leaf Blights :**

Two or 3 foliar sprays with 0.25 to 0.30 mancozeb or zineb at 10-15 days interval may be applied. Application of urea followed by a light irrigation has helped to minimize the disease as this disease sometimes becomes serious under poor management, i.e., under low N and water stress.

**4. Peronosclerospora Downy Mildew :**

Resistant cultivar like suwan, G-2 etc. be used. Planting should be done at least a fortnight before the onset of rains. Burn the crop residue after the harvest. Treating maize seed with a systemic fungicide, metalaxyl @ 4 g/kg as slurry treatment eliminates this group of diseases almost completely.

**5. Stalk Rots :**

Stalk rots are of variable importance from region to region and from season to season. Sometimes, they assume epidemic proportion in some areas depending upon the temperature, rainfall, soil drainage, soil type, fertility level and plant population. The stalk rots usually affect the basal internodes of the plant. As a result the nutrient and water do not reach the above ground parts of the plant. Consequently, the affected plants do not produce ears or even if they produce they are small and chaffy. Very often the affected plants wilt and die prematurely.

**(i) Bacterial stalk rot (*Erwinia chrysanthemi* var. zeae ) :**

Control (i) Cultural practices such as avoiding higher population levels and water logged conditions are helpful in reducing the disease. (ii) Ganga safed-2 hybrid has been observed to be tolerant to this disease (iii) Application of bleaching powder with 33% available chlorine at a concentration of 100-1000 ppm commencing from the stage where the crop is about four week old and repeated at 10-11 days interval, depending upon the severity of the disease, has found to be effective in checking.

**(ii) Pythium stalk rot (*Pythium aphanidermatum*) :**

Control (i) Planting of hybrid Ganga safed-2 or composite Vikram is recommended. (ii) Crop sanitation such as removal of previous crop debris is helpful in checking the disease. Planting between 10th-20th July in the nothern plains has been found to reduce disease intensity. (iii) Apply a solution of captan (N-trichloromethyl-thio-4-cyclohexene 1-2-dicarboximide) or thiram mixed in water @ 2.5 kg/100 l of water/ha on lower internodes of plants 30-35 days after planting. Plant healthy seed treated with fungicide: like thiram @ 2.5 g/kg. Arrange proper drainage and maintain lower population i.e. 50,000 plants/ha.

**(iii) Late wilt and stalk rot (*Cephalosporium maydis*) :**

The disease become apparent usually after flowering as premature wilting of plants. The basal internodes are observed to be discoloured and yellow. These have brown purple or black streaks on the rind. In some cases a wet rot condition is also met with. On splitting the stalk longitudinally the tissue is observed to have become brown. The disease is mainly soil borne and to some extent seed borne also.

**(iv) Black-bundle disease (*Acremonium strictum*) :**

Plant wilt and dry prematurely. The stalks shrink and show a dry rot condition. Long streaks are noticeable on the rind. The vascular bundles become black. This can be seen by cutting the stalk obliquely across when black spots can be seen which represent the affected bundles.

**(v) Charcol rot (*Macrophomina phaseoli*) :**

At flowering time if low moisture plus high temperature conditions prevail the disease manifest itself. The stalk of the infected plants have minute black fungal bodies. The affected stalks become weak resulting in the premature drying and lodging of the plants. On splitting the stalks the pith is observed shredded and the bundle lies in loose strands. The pith and bundles are covered with black fungal sclerotial bodies giving them a grey appearance.

**Control :** Use of Ganga-9, Gang-101, Him 128, Kanchan, Diara Comp, Hunius and Sangam is recommended. The disease can be kept under check by careful water management to avoid low moisture stress conditions. Crop rotation avoiding the natural hosts of the organisms is also effective in reducing the disease. Apply balanced fertilizer doses and avoid high doses or N and low level of ptoash. Use low level of population i.e. upto 50-60 thousand plants/ha.

**6. Ear Rots :**

This disease of maize crop is localised in areas of high rainfall especially in the hilly regions. Economically these diseases are of less importance, hence they have attracted comparatively less attention.

# IV PROBLEMS AND PROSPECTUS OF ATTAINING A MAJOR BREAK THROUGH IN THE MAIZE PRODUCTION

## (a) Problems :

1. Ideal maize plant
2. Cultivars and techniques for different climatic zones i.e. recommendation domain
3. Cultivars and their utilization
4. Herbicides without residual effect
5. Establishment of proper population
6. Reducing volatilization losses of N
7. Miscellaneous-gap between research and farmer.
8. Cultivars for environmental stresses & N-use efficiency.

## (b) Prospects of attaining a major break through :

1. Yield efficiency of tropical maize
2. Shorter maturity in tropical maize
3. Wider adaptations
4. Drought tolerance
5. Crosses between genera (wide crosses): *Tripsacum,* etc.
6. Quality protein improvement
7. Gap between researcher and farmer
8. Possibilities of *Rabi* maize in non conventional areas.

## REFERENCES

Anonymous. (1964-1988) Annual Report, AICMIP, New Delhi.

Anonymous. (1974) Ann. Rep. AICRP Maize, IARI, New Delhi.

Anonymous. (1980) Ann. Rep. AICRP Maize, IARI, New Delhi.

Anonymous. (1981) Ann. Rep. AICRP Maize, IARI, New Delhi.

Anonymous. (1976-1986) CIMMYT Review, Mexico

Anonymous. (1988) Technology for increasing maize production in India. IARI, N. Delhi.

Bhatnagar, GS. (1987) Response of hybrid & composite maize to N at different sowing dates during winter. Ph. D. Thesis, Raj. Ag Univ., Udaipur.

CIMMYT. (1987 & 1986) World Maize Facts and Trends. The economics of Commercial Maize Seed Production in Developing Countries, Mexico DF CIMMYT.

De Leon Carlos. (1984) Maize Diseases. A Guide for field identification. Maize Programmes, CIMMYT.

Gaur, BL. (1991) Indian J.Agron. 36(1): 107-108.

Gupta, OP and Lamba, PS. (1978) Modern Weed Science Today & Tomorrow Printers & Publishers, New Delhi, 421 pp.

Jain, GL. (1986) Role of non-monetary inputs in maize production in India. In Low cost production technology for farmers. Published- IFFCO for Directorate of Extension, Min. of Ag. N.Delhi.

Jaganathan, MT, Morathan, YB and Ramiah, S. (1979) Studeis on the effect of maize and soybean association in different proportions on yield and grains. Madras. agric. J. 66 : 719-721.

Jain, GL, Shah, VH, Dungarwal, HS, Mathur, MK and Singh, K. (1972). Indian J. Agron 17: (3) : 123-127.

Jain, GL. (1973) Indian Fmg. 23 : 17-20.

Jain, GL; Singh, K and Sharma, HN (1973) Indian J. Agron. 20: (1) : 84-85.

Jain, GL. (1974) Indian Farmer's Digrest 6:9-12.

Jain, GL. (1975) World Crops 27(2): 84-90.

Jain, GL, Jangir, RP, Manohar, RS, Dayanand, Singh, SN and Jain, OP. (1981). Indian Soc. Agron. Nat. Symp. Hissar, pp 196-216.

Jain, GL, Chauhan, GS & Manohar, RS. (1983) Agronomic practices for improving N use efficiency in maize. IN: 25 years of Maize Research in India (1957-82) AICMIP, ICAR pp 83-96.

Jain, GL & Mahapatra, IC. (1987) Fertilizer Use for Intensive Maize Production. ICAR Publication.

Jain, GL, Tiwari, RC and Singhi, SM. (1981) IN Crop management to meet the New Challenges. Indian Soc. Agron. Nat. Symp. Hissar, pp 165-169.

Jain, GL. (1981) Indian Fmg. 31(8) : 9-11.

Jain, GL. (1982) Indian Fmg. 32(2) : 5 & 7.

Jain, GL, Singhi, SM, Sharma, HN & Mahnot, SC. (1982). Indian J. Agron. 27: 41-47.

Jain, GL and Jain, KC. (1971) Res. J. Univ. Udaipur 9: 78-79.

Jain, GL, Saroha, MS, Singhi, SM and Jangir, RP. (1983) In: Symp. Agronomy 2000 A.D. Looking Ahead' PRDKVV, Nagpur Feb. 7-9 pp 21.

Jain, GL, Singh, HG, Jangir, RP and Manohar, RS. (1983) In: Proc. Symp. Agronomy 2000 AD- Looking Ahead 'PRDKVV, Nagpur Feb. 7-9 pp 21.

Jain, GL, Singh, RM, Singhi, SM and Jangir, RP. (1983) IN Proc. Symp. Agronomy 2000 A.D. - Looking Ahead " PRDKVV, Nagpur Feb. 7- 9 pp 22.

Jain, GL, Sukharia, DR, and Manohar, RS. (1989) Indian J.Agron. 34(3) : 346-349.

Jain, GL; Sharma, GL and Rao PS. (1987) Response of maize to some new plant growth stimulatns at varying levels of N: IN: Crop Productivity-. Proc. of the Workshop on 'Increasing crop Productivity. HLRF, Bombay, Oxford & IBH Co. Pvt. Ltd, N. Delhi. pp 153-161.

Kalra, GS and Gangawar, B. (1980) Indian J. Agron. 25: 181-185.

Khera, AS, Brar, HS and Sharma, RK. (1979) Indian Jr. Agric. Res. 19(1) : 23-25 (Anonymous, 1982).

Mehts, RS. (1991) Response of maize to N carriers at varying levels of N with Plant Growth Regulators. M.Sc. Ag. Thesis, RAJAU, Udaipur.

Moolani, MK. (1965) Indian J. Agron. 10(3): 313-318.

Moolani, MK and Hukkeri, SB. (1965) Indian J. Agron. 10(1): 45-47.

Naidu, EV and Reddy, MR. (1975) Indian J. Agron. 20(3): 293-94.

Ortega, AC. (1987) Insects nocivos del maize. Una guia para su identification en el campo Mexico D.F. CYMMYT.

Rao DS. (1985) Nitrogen and Phosphorus management in maize intercropped with *mung* and cowpea. M.Sc. Ag. Thesis, Sukhadia University, Udaipur.

Rao, DS, Jain, GL and Sharma, GL. (1987) Response of intercropped mungbean/cowpea to N and P applied to maize and their residual effect in wheat. IN: 2nd Int. Symp. On Mungbean, Bangkok, Thailand,(Nov. 16-21) Presented.

Rathore, DN, Virbal Singh and Singh, BP. (1976) Indian J. agric, Res. 10: 79-82.

Reddy, HA; Reddy, KH and Reddy, MB, (1980) Experimental Agriculture. 16(2): 179-184.

Sandhu, KS and Gill, GS. (1973) Indian J.Weed Sci. 5: 1-5.

Shah, MH, Kaul, PK, Khanday, BA and Kachroo, D. (1991) Indian J. Agron. 56(1): 23-28.

Sharma, HC and Upadhyaya, UC. (1973) Indian J. Agron. 18(2) : 220- 221.

Sharma, RN. (1978) Indian J. Agron. 23(3) : 248.

Shukla, SP, Kalia, BD and Modgal, SC. (1978) Indian J. Agron. 23 (3): 185.

Singh, HG. (1968) Science & Culture, 34: 305-306.

Singh, J. (1980) Breeding Production and Protection Methodologies of maize in India, AICMIP, ICAR, N. Delhi.

Thakur, Dr. Singh, KK, and Thakur, RC. (1990) India J. Agron. 35(4): 357-360.

Tiwana, and Puri. (1976) Indian J.Agron. 21(2): 75-76.

Ullstrup, AJ. (1966) Corn diseases in United States and their control. USDA Agric. Hanbk. No. 199 : 3.

Dryland Resources & Technology, 7 (1992)
Editors : L.L. SOMANI, R.L. SHYAMPURA & M.C. MUNDRA


# Reclaiming Sodic Soils Through Waste and Byproduct Sulphuric Acid

*L.L. Somani*[*] *K.L. Totawat*[*]
*K.K. Vyas*[**] *I.J. Gulati*[*] *& K.C. Laddha*[**]
[*] Rajasthan College of Agriculture, Udaipur
[**] SKN College of Agriculture : Jobner

## ABSTRACT

A critical examination of the results of several laboratory and field investigations involving use of industrial waste and byproduct sulphuric acid show that it could be effectively used for reclaiming alkali soils. Sulphuric acid is known to supply soluble calcium indirectly through dissolution of insoluble $CaCO_3$ for replacement of sodium from the exchange complex of alkali soils, besides facilitating the removal of soluble sodium salts through leaching. Soil physical properties like hydraulic conductivity, water stable aggregation, infiltration, etc. are also improved following treatment of sodic soil with $H_2SO_4$. The fertility status of treated soil is considerably improved. Use of waste and by-product $H_2SO_4$ for ameliorating alkali soils, particularly if such soils happen to be close to the industries producing it, is an attractive preposition because surface spreading of waste $H_2SO_4$ will not only help in alkali amelioration but also in decreasing pollution.

## INTRODUCTION

Pollution due to industrial waste is such a complex and persistent problem that continued concern and action will be needed throughout the forseeable future. Using soil as a depollutant is an extension of natural recycling and can be a highly beneficial way to dispose off most organic and other wastes. Materials that would be pollutants anywhere else become energy source for soil microorganisms, nutrient source for plant growth or serve as an amendent for alkali amelioration when applied to the soil.

Considerable amount of sulpuric acid (waste and by-product from dairies, smelters, petroleum industries) is being released every year which is polluting the environment. An attempt has been made in the present article to examine the possibilities of using this waste for reclaiming alkali soils.

## WASTE AND BY-PRODUCT $H_2SO_4$

The potential value of sulphuric acid for improving the productivity of sodic soils has long been recognised. However its actual use has not been extensive, partly because of limited supplies, handling difficulties, inadequate application criteria and partly because of high cost. Production of by-product $H_2SO_4$ is greatly increasing in recent past which is expanding the problem of environmental pollution (Miyamoto *et al.* 1975). This would eventually necessitate proper disposal and management of waste $H_2SO_4$. Agricultural use of by-product and waste $H_2SO_4$ may offer a partial solution. An attempt has been made here to review and explore such potential uses. Special attention is given to amelioration of sodic soil which is likely to absorb quantities of waste surplus acids.

## POTENTIAL SOURCES OF $H_2SO_4$

The production of waste and by-product $H_2SO_4$ is likely to be higher and higher as the air pollution abatement are implemented in smelters, coal burning power plant, sulphur burning plants, petroleum and dairy industries.

Considerable amount of $H_2SO_4$ is released as waste at milk collection centres of various dairies after evaluating the fat content of milk samples. This waste $H_2SO_4$ could be collected in plastic carbuoys and applied to sodic soils. This waste from dairy industry contains about 40% $H_2SO_4$ in addition to acid mixture of milk.

Copper, zinc and lead smelters usually recover less than one-fourth of the sulphur and rest is mostly discharged as $SO_2$. Considerable $H_2SO_4$ will be produced if air pollution regulations are fully implemented. Most of these smelters are capable of producing over 1000 tonnes of concentrated (90-95%) sulphuric acid per day (Miyamoto *et al.* 1975). The coal burning power plants and petroleum industries can yield millions of tons of $H_2SO_4$ annually.

## CHEMISTRY OF ALKALI AMELIORATION THROUGH $H_2SO_4$

Successful reclamation of sodic soils by gypsum, pyrites or elemental S has been reported in numerous publications (Dargan *et al.* 1982; Pahwa and Gupta, 1982; Mehta, 1983). Information concerning the reclamation of sodic soil through by-product and waste $H_2SO_4$ is however limited. As early as 1915, Hunt (1916) was successful in reclaiming alkali soil by surface sprinkling of $H_2SO_4$. Concentrated $H_2SO_4$ sprinkled on surface of sodic soils at a rate of 1.5 t/ha increased water infiltration rates and seedling emergence and resulted in good growth of crops (Mc George *et al.*, 1956) indicating promising prospects of waste $H_2SO_4$ in alkali amelioration. Concentrated $H_2SO_4$ sprinkled at the rate of 3-5 t/ha with leaching successfully reclaimed sodic soils (Miyamoto and Stroehlein, 1975).

The reclamation of alkali soils with $H_2SO_4$ is based upon the following chemical reactions.

If the soil is calcareous :

$$H_2SO_4 + CaCO_3 \longrightarrow CaSO_4 + CO_2 + H_2O$$

$$2Na\text{-}Clay + CaSO_4 \longrightarrow Ca\text{-}Clay + Na_2SO_4$$

but if the soil is non-calcareous:

$$2Na\text{-}Clay + H_2SO_4 \longrightarrow H\text{-}Clay + Na_2SO_4$$

The ameliorative effect of $H_2SO_4$ is very quick particularly if the soil is calcareous. It brings about instantaneous improvement in infiltration and other physical properties due to flocculation of highly dispersed clay. Sharma et al. (1971) reported the superiority of $H_2SO_4$ over $CaSO_4$ in improving percentage pore space; permeability, stability index and aggregates greater than 0.25 mm. Tyulenina and Fomina (1978) observed increased activity of soil microflora, decreased peptization of clay and increased available moisture reserve following application of $H_2SO_4$. Sulphuric acid application increased the rate of water penetration for most sodic soils, the effect being greater with increasing levels of exchangeable Na (Yahiya *et al.*, 1975). Kurbaton *et al.* (1975) reported drop in electrokinetic potential of the şolonetzes from 28 to 13 mV thus aggregating the soil particles and increasing the average radius of the pores and the hydraulic conductivity of soils. Pertosian and Avundjian (1975) reported that acidification with $H_2SO_4$ affected neutralization of sodicity and decreased the ESP significantly. When the soil is being acidified with $H_2SO_4$, the $Ca^{++}$ of the newly formed gypsum replaces the exchangeable sodium. The hydrophilic properties of soils are improved. Maximum hygroscopicity of sodic soil before and after $H_2SO_4$ treatment varied from 8 to 17 and 4 to 7.8 per cent, respectively. Water holding capacity also increased from 26 to 35 per cent.

Chauhan and Tripathi (1980), however reported harmful effect of $H_2SO_4$ in non-calcareous sodic soil. Bocskai (1969) stressed the use of waste $H_2SO_4$ in conjunction with lime dust for non- calcareous alkali soils of Hungary.

The high pH and lime content of sodic soils often limits solubility and availability of phosphours and certain micronutrients like Fe, Mn and Zn. In general the acidification of calcareous soils increases the solubility and availability of these elements originally present in soils and those applied as fertilizers. $H_2SO_4$ applied to P-deficient calcareous soils increased the water and $CO_2$ saturated water solubilized P severalfold with the maximum release at acid rates sufficient to neutralize 40-60% of acid titrable basicity (Miyamoto and Stroehlein, 1975; Ryan and Stroehlein, 1973). Water soluble and DTPA extractable Fe, Mn and Zn increased markedly following $H_2SO_4$ treatment (Ryan *et al.*, 1974).

## LABORATORY AND FIELD INVESTIGATIONS WITH WASTE AND BY-PRODUCT $H_2SO_4$

Sharma *et al.* (1983) in their laboratory studies using soil columns, found addition of dairy waste $H_2SO_4$ having 0.5 to 1.0 per cent $H_2SO_4$ in diluted form to be very effective in increasing the water intake rate of highly sodic soil. The pH of the surface (0-15 cm) soil decreased from 10.3 in original soil to 8.1 and 7.7 when this waste was applied at the rate of 50% GR having 0.5 and 1.0% $H_2SO_4$

concentration, respectively. They recommended that such dairy waste, after diluting, could be effectively utilized in reclaiming alkali soils in the vicinity of dairies and milk collection centres. Shah (1975, 1977) made field scale utilization of waste $H_2SO_4$ for reclaiming alkali land in Gujarat State. His results showed that the soil pH dropped from 9.2 to 7.2 and the hydraulic conductivity improved 20 times.

Kurbaton *et al.* (1975) reported increased yield of barley and oat hay on a soda solonetzes of Western Siberia following application of industrial waste $H_2SO_4$. Palkhiwala and Shah (1975) suggested the use of petroleum acid resins in reclaiming the saline-sodic soils of Saurashtra. Bocskai (1969) found petroleum resins, (obtained as a by-product during the sulphuric acid refining of coal petroleum) to be a promising product for reclaimation of calcareous alkali soils and in conjunction with lime dust for non-calcareous alkali soils of Hungary. The petroleum resins consisted of large molecule of organic substances as well as, 10-40 per cent $H_2SO_4$ depending upon refining procedure.

For efficient amelioration of sodic solonetz-solonchaks soils, application of $H_2SO_4$ should be followed by leaching to neutralize the alkalinity of the soil solution and displace exchangeable sodium from the upper one meter of the soil. To increase the effectiveness of chemical improvement, the soil must be diked during acidification after it is graded and loosened (Petrosian, 1978). In the zone of penetration, sulphuric acid decomposes calcium and magnesium carbonates giving rise to the formation of soluble calcium sulphate which positively affect soil properties. The newly formed Ca and Mg salts participate in the soil colloidal chemical reactions replacing active/toxic sodium from the exchange complex (Petrosian and Tchitchian, 1969). Petrosian (1977) observed the ameliorative influence of waste $H_2SO_4$ in a very short time. High yielding annuals and perennial crops can be successfully cultivated on treated sodic soil, securing very quick returns on capital investments.

## WASTE AND BY-PRODUCT $H_2SO_4$ VIS-A-VIS TRADITIONAL AMENDMENTS

The relative performance of $H_2SO_4$, sulphur, gypsum, etc. has been evaluated by a number of workers under field conditions. Chena (1959) reported that $H_2SO_4$ removed the Na almost twice as rapidly as gypsum did. Dolozhenko (1977) evaluated the effect of equivalent amount of $H_2SO_4$, $HNO_3$, $CaCl_2$ and $CaSO_4$ on the permeability and Na replacement from sodic solonchaks in West Siberia. The results of his field study indicated that the ameliorative effect of $HNO_3$, $CaCl_2$ and $CaSO_4$ was approximately same during prolonged leaching but was much weaker than the effect of $H_2SO_4$. Branson and Fireman (1960) studied the possibility of reclaiming a saline-sodic soil in Palo Verde Valley, Californea using several amendments. Sulphuric acid was found to be more effective than gypsum. Gypsum treatment needed a much extended period of leaching to increase crop yields to the same extent as the $H_2SO_4$ did. Experiments of Berezin *et al.* (1978) concluded in the forest-steppe zone of the Omsk province (USSR) showed the feasibility of using concentrated $H_2SO_4$ solution for improving the physico-chemical properties of a solonetz with a high Na content and $NaCO_3$ salinization. The advantage of soil acidification lies in the accelerated displacement of the exhangeable Na during the first two years where as the benefits of gypsum application become apparent in forth year in the more complete saturation of exchange complex by calcium. Agabayan and Rafaelyan (1963) and Alawi *et al.* (1980) also reported that when applied on chemical equivalent basis, soil acidification with $H_2SO_4$ proved superior to gypsum application.

Overstreet *et al.* (1951) carried out a field trial on a 100% Na-saturated soil using S, $CaSO_4$ and $H_2SO_4$ (93%) at chemically equivalent rates. The yields of unirrigated pasture established after treatment were significantly higher in plots receiving $H_2SO_4$ than in those with $CaSO_4$ and S. Analysis of surface soil showed that $H_2SO_4$ greatly reduced pH and Na values of the soil (ESP from 60 to 10). The $H_2SO_4$ obviously reacts very quickly with alkali soils. Poonia and Bhumbla (1974) compared the efficiency of three lime solubilizing amendments, viz. $H_2SO_4$, HCl and $Al_2(SO_4)_3$ in reclaiming a saline-alkali soil. Calcium content in the plant increased significantly with the application of these amendments and it was in the decreasing order : $H_2SO_4$, $Al_2(SO_4)_3$, HCl, indicating the superiority of $H_2SO_4$.

Table 1 : Effect of gypsum and sulphuric acid treatment on grain yield of rice and wheat (g/ha$^{-1}$)

| Treatments | 1969–70 | | 1970–71 | | 1971–72 | |
|---|---|---|---|---|---|---|
| | Paddy | Wheat | Paddy | Wheat | Paddy | Wheat |
| Control | 15.1 | 7.2 | 21.2 | 17.2 | 35.6 | 14.6 |
| Gypsum at 80% GR | 36.7 | 19.2 | 43.6 | 26.1 | 51.3 | 27.4 |
| $H_2SO_4$ at 80% GR | 45.6 | 25.6 | 46.8 | 31.0 | 55.3 | 30.5 |

Source : Agarwal *et al.* (1979)

Prather *et al.* (1978) used three amendments, viz $CaSO_4$, $CaCl_2$ and $H_2SO_4$ singly and in combination to test their effectiveness and efficiencies with respect to amount of amendment, time and leaching needed. As a single amendment, $H_2SO_4$ was more effective than $CaSO_4$. Combining $H_2SO_4$ with $CaSO_4$ appreciably reduced the time and leaching needed to achieve reclamation as compared with $CaSO_4$ alone. The benefits in terms of crop yield due to application of $H_2SO_4$ on a saline-alkalli soil (initial pH 9.1 to 9.8, EC 4 to 6 mmhos/cm, ESP 41 to 60) over a period of three years (Table 1) show remarkable superiority of $H_2SO_4$ over gypsum (Agarwal *et al.* 1979). At time the use of $H_2SO_4$ on equivalent basis is slightly costlier than gypsum. The quicker reclamation with $H_2SO_4$, however, pay off more than higher initial cost by harvesting good yields following $H_2SO_4$ treatment (Kovda *et al.* 1973). Huge availability of waste and by-product $H_2SO_4$ offers a good alternative to commercial $H_2SO_4$.

Reclamation with gypsum, sulphur or pyrites, etc, is a very slow process and proceeds gradually with time through small depths. They largely bring about reclamation of the surface layer of alkali soils. As a result, the entire profile is not reclaimed as is the case with $H_2SO_4$. Even after applying the traditional amendments (gypsum, pyrites or sulphur) the farmer has to grow tolerant and semi- tolerant crops for several years, but can go for normal cultivation to reap good profits with $H_2SO_4$ treatment. Being corrosive and hazardous ($H_2SO_4$ severely attacks the skin and clothes of the user upon contact), its use requires proper equipment and safety training for handling and applications.

## FIELD APPLICATION OF $H_2SO_4$

Sulphuric acid is usually sprinkled in concentrated form on soil surface under gravity flow by tractor or truck equipped with sprinklers (Miyamoto and Stroehlein, 1975). When acid application rates are less than a few metric tons per hectare, it can be applied through leaching water or the first irrigation. Water applied acid at high rates, however, causes corrosion of irrigation systems, destroys soil aggregates, impairs the penetration of leaching water by entrapping $CO_2$ and may result in poor distribution over a large field (Miyamoto *et al.* 1974).

Some commercial operators in USA and USSR have used a special acid injection rig attached to a tractor (Tisdale, 1970). Essential features consist of a steel or stainless steel tank and knifing edges with injection pipes. Such a rig can reportedly inject acid at about 30 hectares per day. This type of custom application not only eliminates large investment by each farmer, but also reduces the potential hazard involved in acid handling and application.

## CONCLUSION

i.) Waste $H_2SO_4$ is effective in ameliorating calcareous alkali soils and non-calcareous alkali soils in conjunction with lime.

ii.) Waste $H_2SO_4$ brings about instantaneous improvement in soil physical properties besides decreasing the ESP of sodic soils.

iii) Surface spreading of waste and by-product $H_2SO_4$ not only reclaims sodic soils but also helps in reducing environmental pollution.

iv) The use of $H_2SO_4$ requires proper equipment and safety training for handling and application.

## REFERENCES

Agababyan, V.G. and Rafaelyan, A.S. 1963. The dynamics of allkalinity in improved soda solonchaks. *Tr. nauchnoissled. Inst. Pochvoved. Agrokim. Armenian SSR* 2: 211-223.

Agarwal, R.R., Yadav, J.S.P. and Gupta, R.N. 1979. Saline and Alkali Soils of India. ICAR, New Delhi p. 286.

Alawi, B.J., Stroehlein, J.L., Hanlon, E.A. and Turner, F. 1980. Quality of irrigation water and effects of sulfuric acid and gypsum on soil properties and sudangrass yields. *Soil Sci.* 129 : 315-319.

Berezin, L.; Voropaeva, Z. and Kurbatov, A. 1978. The effect of gypsum application and soil acidification on the physico-chemical properties of a crusted solonetz *Nauchn. Tr. Omsk. S-Kh. Inst.* 171 : 3-8.

Bocskai, J. 1969. The use of acid resins supplied by the oil industry in the chemical amelioration of alkali soils. *Agrokem. Talajt.* 18(Suppl) : 336-338.

Branson, R.L. and Fireman, M. 1960. Reclamation of an impossible alkali soil. *Trans. 7th int. Congr. Soil Sci.* 1 : 543-552.

Chauhan, C.P.S. and Tripathi, B.R. 1980. Influence of continuous and intermittent leachings with and without amendments on reclamation of saline-sodic soils. *Madras agric. J.* 67 : 109-112.

Chena, G. 1959. Reclamation of saline sodium soil. *Agric. Tech. Mexico.* 8 : 17-19.

Dargan, K.S., Singh, O.P. and Gupta, I.C. 1982. Crop production in salt-affected soils. Oxford & IBH Publishing Co., New Delhi, pp. 276.

Dolozhenko, I.B. 1977. Effectiveness of chemical amendments in the leaching of sodic solonchaks in West Siberia. *Pochvovedenie* 2 : 64-72.

Hunt, T.F. 1916. Ann. Rept. Calif. Agr. Expt. Sta. Californea p.100.

Kovda, V.A.; Berg, C.V.D. and Hagan, R.M. (Ed) 1973. Irrigation, drainage and salinity. An International source book. Hutchinson / FAO / UNESCO, London, p. 510.

Kurabation, A.I.; Okorokov, V.V. and Berezin, L.V. 1975. The effect of amendments on the electrokinetic properties of solonetzes of Western Siberia. Izv. Timiryazev *S-kh Akad.* 5 : 104-108.

McGeorge, W.T., Breazeale, E.I. and Abbott, J.L. 1956. Polysulfides as soil conditioners. *Ariz. Agric. Exp. Sta. Tech. Bull.* 131

Mehta, K.K. 1983. Reclamation of Alkali soils in India. Oxford & IBH Publishing Co., New Delhi, p. 280.

Miyamoto, S. and Stroehlein, J.L. 1975. Sulfuric acid for increasing water penetration into some Arizona soils. *Prog. Agric. Ariz.* 27 : 13-16.

Miyamoto, S., Bohn, H.L., Ryan, J. and Yee, M.S. 1974. Effect of sulfuric acid and sulfur dioxide on the aggregate stability of calcareous soils. *Soil Sci.* 118 : 299-303.

Miyamoto, S., Ryan, J. and Stroehlein, J.L. 1975. Potentially beneficial use of sulfuric acid in South-western agriculture. *J. Environ Qual.* 4 : 431-437.

Overstreet, R.; Martin, J.C. and King, H.M. 1951. Gypsum, sulfur and sulfuric acid for reclaiming an alkali soil of the Fresno series. *Hilgardia* 21 : 113-127.

Pahwa, K.N. and Gupta, I.C. 1982. Reclamation and management of salt-affected soils. Associated Publishing Co., New Delhi, pp. 352.

Petrosian, G.P. 1977. Chemical reclamtion of solonetz-solonchaks in the Ararat plain and the possiblity of utilizing saline waters for leaching and irrigation. *In* Managing Saline Waters for irrigation. Ed by H.E. Dregne. *Texas Tech. Univ.*, Lubbock, Texas, p. 466-479.

Petrosian, G.P. 1978. Technology and economic indexes of the chemical improvement of the sodic solonetz-solonchaks of the Ararat plain of the Armenia SSR. *Pochvovedenie* 9: 59-73.

Petrosian, G.P. and Avundjian, Z.S. 1975. Chemical reclamation effect on physical and water-physical properties of soda saline soils in pre-Arax low land. *Int. Symp. New Develop Field Salt Affected Soils, Cairo, 1972.* p. 513-523.

Petrosian, G.P. and Tchitchian, A.I. 1969. Soda-saline soils of the Ararat plain and methods for their reclamation. *Agrokem. Talajt.* 18 (suppl) : 121-140.

Poonia, S.R. and Bhumbla, D.R. 1974. Effect of $H_2SO_4$, HCl and $Al_2(SO_4)_3$ on the yield, chemical composition and Ca uptake from applied $Ca^{45}CO_3$ by dhaincha (*Sesbania aculeata* Pers) in saline-alkali soil. *Pl. Soil* 40: 557-564.

Prather, R.J., Goertzen, J.O.; Rhoades, J.D. and Frenkel, H. 1978. Efficient amendment use in sodic soil reclamation. *Soil Sci. Soc. Am. J.*, 42 : 782-786.

Ryan, J. and Stroehlein, J.L. 1973. Use of sulfuric acid on phosphorus deficient Arizona soils. *Prog. Agric. Ariz.* 25 : 11-13.

Ryan, J., Miyamoto, S. and Stroehlein, J.L. 1974. Solubility of manganese, iron and zinc as affected by application of sulfuric acid to calcareous soils. *Pl. Soil* 40 : 421-427.

Shah, R.K. 1977. A quick method for reclaiming saline-alkali soil using dairy waste sulphuric acid. *Agril. Res. News. I.* 3 : 4-6.

Shah, R.K. 1977. A quick method for saline -alkali soil reclamation using sulphuric acid. Proc. *Indo-Hungarian Sem. on Management of Salt-Affected Soils.* CSSRI, Karnal, pp. 134- 139.

Sharma, D.P.; Mehta, K.K. and Singh, G. 1983. Use of dairy waste for the reclamation of sodic soils. *Curr. Agric.* 7 : 13- 17.

Sharma, V.K.; Mehrotra, C.L. and Saxena, V.P. 1971. Effect of amendments on the physical conditions of saline-alkali soils. *All India Symp. Soil Salinity, Kanpur,* p. 141-144.

Tisdale, S.L. 1970. The use of sulfur compounds in irrigated acid-land agriculture. *Sulphur Inst. J.* 6 : 2-7.

Tyulenina, L. and Fomina, V. 1978. Chemical improvement of solonetzes by the method of aerosol acidification. *Nauchn. Tr. Omsk. Skh. Inst.*, 171 : 8-14.

Yahia, T.A., Miyamoto, S. and Stroehelin, J.L. 1975. Effect of surface applied sulfuric acid on water penetration into dry calcareous and sodic soils. *Soil Sci. Soc. Am. Proc.* 39 : 1201-1204.

Dryland Resources & Technology, 7 (1992)
Editors : L.L. SOMANI, R.L. SHYAMPURA
& M.C. MUNDRA


# Diseases of Dryland Oilseed Crops in India

*S.R.S. Dange, R.N. Pandey*
*& B.G. Desai*
Department of Plant Pathology,
College of Agriculture, GAU,
Sardar Krushinagar–385 506

## ABSTRACT

Peanut, Sesame, Sunflower and Niger are important oilseed crops of dryland agriculture, cultivated in about 10.9 million hectares and producing about 7.1 million tonnes of oilseed annually in India. Oilseed crops are affected by many fungal, bacterial, viral and nematode diseases which are responsible for considerable yield losses. Most of the released varieties at present do not possess adequate degree of resistance against major diseases, though the resistance has now been identified for most of them. Plant protection measures to manage important diseases of these crops are playing an important role in increasing oilseed production.

Seed treatment with suitable fungicides like thiram, captan, carbendazim, mancozeb etc. is very effective in controlling seed rot and seedling diseases particularly in groundnut. Majority of farmers do not adopt even such a simple seed treatment which is a must to obtain proper plant stand for a good harvest.

Apart from seed rot and seedling diseases, groundnut suffers with leaf spots, rust and other minor fungal diseases. The leaf spots and rust can be effectively managed by spraying 0.2% chlorothalonil or mancozeb. Besides fungal diseases, viral, nematode, mineral deficiency and aflatoxin problems are also of much concern. Sesame crop too is affected by many fungal, bacterial, viral and mycoplasma diseases. The Alternaria leaf spot can be managed by spraying 1% Bordeaux mixture, or 0.1% zineb. White leaf spot and angular leaf spot caused by two different species of *Cercospora* are manageable by carbendazim or *Bayleton.* Phytophthora blight, powdery mildew, Fusarium wilt, charcoal rot are other important diseases which are checked effectively by adopting suitable control measures. Bacterial blight and leaf spot are kept under check by spraying *Agrimycin.* Phyllody is caused by MLOs which is known to spread through leaf hopper (*Orosius albicinctus*). Application of Thimet 10G or Phorate 10G or Temik 10G in soil at sowing time combined with spraying of 0.1% Metasystox satisfactorily check the spread of the disease.

Besides seed rot, seedling diseases, and other minor fungal diseases of sunflower, Alternaria blight, downy mildew, rust, powdery mildew, charcoal rot, *Sclerotinia* root rot or wilt are major ones. The variety SC 17 proved highly resistant to bacterial leaf spot disease. Sunflower mosaic, yellow ring mosaic and yellow spot are important viral diseases. Nematode disease and phanerogamic parasite *Orobanche* are common problems on this crop.

In niger, collar rot (*Corticium rolfsii*), leaf spots caused by species of *Alternaria, Cercospora* and *Curvularia,* powdery mildew etc. are a few fungal diseases. Bacterial leaf spot disease and phanerogamic parasite *Cuscuta* have also been observed on this crop. Measures to manage diseases of these oilseed crops have been discussed. Plant protection measures, if adopted as package of practices in cultivation of oilseed crops, will play an important role in minimising the yield losses to a great extent.

## INTRODUCTION

Commercially cultivated important dryland annual oilseed crops in India are peanut, sesame, sunflower and niger which serve as raw material for vegetable oils. Presently, these crops are cultivated in 10.9 million hectare, producing about 7.1 million tonnes of oilseeds. The area, production, productivity number of varieties released and their oil content are given in Table -1.

About 13.5 kg of oil/fat should be made available per capita per annum, but present availability is hardly 6.5 kg of visible and 2 kg of invisible oil/fat, and there is shortage of 5 kg per capita per annum in India. Fat/oil requirement of different categories of

Table 1 : Area, production, productivity (1986–87)*, oil content and varieties released of different dryland oilseed crops in Indian.

| Crop | Area (1000 heactares) | Production (1000 tonnes) | Productivity (kg/ha) | Per cent of the total | | Oil (%) | Varieties released |
|---|---|---|---|---|---|---|---|
| | | | | Area | Production | | |
| 1. Peanut | 7,150.9 | 6,059.5 | 847 | 65.5 | 85.7 | 44–50 | 55 |
| 2. Sesamum | 2,183.2 | 443.2 | 203 | 20.0 | 6.3 | 46–52 | 35 |
| 3. Sunflower | 992.1 | 435.7 | 439 | 9.1 | 6.2 | 40–48 | 7 |
| 4. Niger | 588.7 | 132.1 | 224 | 5.4 | 1.8 | 38–50 | 7 |
| | 10,914.9 | 7,070.5 | | | | | |

* Source : Agriculture situation in india, May 1988 (2) : 186–191.

people in a balanced diet for vegetarians is given in Table-2. If edible oil problem is not solved, it will continue to incur heavy expenditure at the cost of several developmental activities of other spheres to fill in the gap of demand and supply of edible oils in India.

Table 2 : Fat and oil requirement* in a balanced diet for vegetarians

| Category | gm/Capacity/day | Remarks |
|---|---|---|
| Adult man | 40 | Moderate work |
| | 50 | Heavy work |
| Adult woman | 35 | Moderate work |
| | 40 | Heavy work |
| Adolescent | 35 | 13–15 year boy |
| | 45 | 16–18 year boy |
| | 35 | 13–18 year girl |
| Children | 20 | 1–3 years |
| | 25 | 4–6 years |
| | 30 | 7–9 years |
| | 35 | 10–12 years |

* Source : Singh, R.P. and Satpathy, C. (1981). Oilseed production in India. Seeds and farms 7(8) : 25-29.

To cope up with demand of oils, the National Commission on Agriculture (1976) has made projections on the area to be covered and the yield to be achieved of dryland oilseed crops (Table-3).

Table 3. Projection of area and the yield of different dryland oilseed crops in India*

| Crop | Area (Million hectares) | | Yield (q/ha) | |
|---|---|---|---|---|
| | Existing | 2000 AD | Existing | 2000 AD |
| Peanut | 7.23 | 9.0 | 7.8 | 15.0 |
| Sesame | 2.38 | 3.0 | 2.1 | 6.0 |
| Sunflower | – | 2.0 | – | 10.0 |
| Niger | 0.48 | 1.5 | 2.4 | 5.0 |

* Source : National Commission on Agriculture (1976)

## I Peanut (*Arachis hypogaea* L.)

Through planting to harvesting peanut suffers with many diseases. Out of several fungi, bacteria, viruses and nematodes involved, fungal pathogens also cause severe pod and kernel rots.

## FUNGAL DISEASES

### 1. Seed Rots and Seedling Diseases

The soil inhabiting or seedborne fungi attacking kernels soon after planting invariably give rise to poor seedling stand due to seed rots and seedling diseases. Commonly occurring pathogens involved in these diseases are: *Aspergillus flavus* Link ex. Fr., *A. niger* van Tieghem, *Botryodiplodia theobrome* Pat., *Diplodia, Fusarium, Macrophomina phaseolina* (Tassi) Goid, *Mucor, Penicillium, Pythium, Rhizoctonia solani* Kuhn, *Sclerotium rolfsii* Sacc., *Rhizopus arrhizus* Fischer, *R. stolonifer* (Ehranb. ex. Fr.) Lind. etc. *A. flavus* and *A. niger* are responsible for pre-emergence and post-emergence death of seedlings apart from rotting. *A. niger* induces collar rot or crown rot which is serious in sandy and sandy loam soils, is deadly at the seedling stage (Dange and Saradava, 1986 ; 1988). Water logging conditions accentuate the problem of rots and seedling diseases. *A. flavus* is also responsible to produce aflaroot disease in seedlings.

The peanut genotypes differ in their reactions to seed rots and seedling diseases (Dange and Saradava, 1985) and the concentrations of sucrose and other chemicals also influence mode of fungal invasion (Ghewande *et al.*, 1982 b). These days Captan or thiram or carbendazim at 2 to 4 g/kg seeds are much in use as effective seed dressers (Dange and Saradava, 1985). Disallowing water logging conditions in the fields is another potent safty measure against seed rots.

As these diseases are easily managed by seed dressers, efforts have not been made to develop resistant varieties.

### 2. Early and Late Leaf Spots

Early and late leaf spot diseases together on peanuts are popularly known as 'Tikka' disease. Early leaf spot is caused by

*Cercospora arachidicola* Hori (perfect stage : *Mycosphaerella arachidis* Deighton) while the late leaf spot is due to *Cercosporidium personatum* (Berk. and Curt.) Deighton which is now designated as *Phaeoisariopsis personata* (Berk. and Curt.) v. Arx (Perfect stage : *Mycosphaerella berkeleyii* W.A. Jenkins). The perfect stages of both organisms are rarely observed in field conditions (Mc Donald *et al.*, 1985).

Due to these leaf spots diseases, in India 15 to 50% crop losses have been estimated (Chohan, 1974). The losses are much higher when leaf spot and rust diseases occur together (Ghewande *et al.*, 1983).

Early leaf spot symptoms appear after 3-4 weeks of planting on lower leaves as clean pale areas gradually turning yellow and finally becoming reddish-brown to black bearing sporulating fungus comprising of conidiophores and conidia on the upper leaf surfaces. Such circular to irregular spots measuring 1-10 mm in diameter are typically surrounded by yellow halos. On lower surfaces, these spots exhibit orange colouration.

Late leaf spot symptoms generally appear after 6-8 weeks from planting as scattered, round, blackish-brown coloured spots on both the surfaces of infected leaves which are mainly surrounded by yellow halos. The spots measuring 1.5 to 5 mm in diameter appear carbon black due to heavy sporulation on their lower surfaces. Late leaf spot is more damaging on account of heavy defoliation, which imparts a characteristic barren look of stumped plants. Variation in types of leaf spot symptoms is not uncommon (Subrahmanyam *et al.*, 1982 a).

The sources of resistance in the screened germplasm are of unstable type against these leaf spot pathogens known for their variability (Subrahmanyam *et al.*, 1985). The work on this line is in progress to tap down sources of resistance (Rathnaswamy *et al.*, 1986). Though the pathogen survives in plant debris, but perithecia do not form in Indian conditions.

Spraying of nonsystemics such as Bordeaux mixture (2:2:50 or 5:5:50) or copper oxide (0.15 to 0.35%) or copper oxychloride (0.2%) or sulphur dusting (15 to 22 kg/ha) proved effective (Kolte, 1985). Spraying of 0.2% wettable sulphur (Tandon and Singh, 1968) suppresses leaf spots. Either of the sprayings of 0.2% mancozeb, maneb and zineb have been promising (Singh and Kang, 1981).

Spraying systemics such as Baycor (750 ml/ha), chlorothalonil (0.2%), triadimefon are equally effective (Dange *et al.*, 1981; Godara and Thakore, 1987). Benomyl, carbendazim and thiophanate methyl at concentrations of 0.02 to 0.05% and 3 application of 0.1% Duter (Fentin hydroxide) control these diseases effectively (Tiwari *et al.*, 1987). At times, combinations of tridemorph + carbendazim or carbendazim (0.05%) + mancozeb (0.2%) have successfully controlled when rust and leaf spots occurring together (Shekhawat *et al.*, 1987; Vyas *et al.*, 1986). Systemic fungicide biloxazole has been found effective against late leaf spot as well as rust also (Thakore *et al.*, 1987). Organo-tin compounds viz. Brestan (0.1%) and Duter (0.1 to 0.2%) have proved better spray fungicides than copper, dithiocarbamate and sulphur fungicides (Lewin and Natrajan, 1971).

### 3. Rust

In 1969, the peanut rust caused by *Puccinia arachidis* Speg. was first observed in the Punjab state of India. Now it has gained much importance in all peanut growing tracts (ICRISAT, 1987). Depending upon variety, rust severity and crop season, losses in pod yield range from 10 to 50% (Ghuge *et al.*, 1981). Also, oil content decreases from 7 to 10% (Kenjale *et al.*, 1981).

Initially orange-red to brown coloured uredopustules measuring 0.2 to 0.5 mm in diameter are visible on the abaxial surfaces of leaves while corresponding adaxial surface look chlorotic. Under severe conditions of disease, the uredopustules coalesce and fresh ones develop on adaxial surface as well. Heavily infected leaves curl and fall down. The occurrence of teliospores of *P. arachidis* within uredia of *A. hypogaea* (Cultivar 'Tatu') under artificial inoculation conditions in a green house studies has been reported from Brazil (Hennen *et al.*, 1976). Further attempts to search for teliospores in rust infected germplasm under Indian conditions have failed (Dange and Desai, 1985), after the discovery of teliospores in India by Chahal and Chohan (1971). Since the pathogen exclusively forms uredial stage in nature, obviously the uredospores are the only means of carry over and spread of the disease in India (ICRISAT, 1987).

In Punjab, the rust is not a problem these days, but in southern states and Gujarat state of India, it is a menace to peanut crop. Abundant uredospores as a potential source of inoculum is readily available for spread and carry over of rust pathogen throughout the year round on account of availability of peanut plants during all crop seasons in a year (Fig.1). In northern states of India, the pathogen fails to survive in the absence of peanut plants in the off season (Sokhi and Jhooty, 1986).

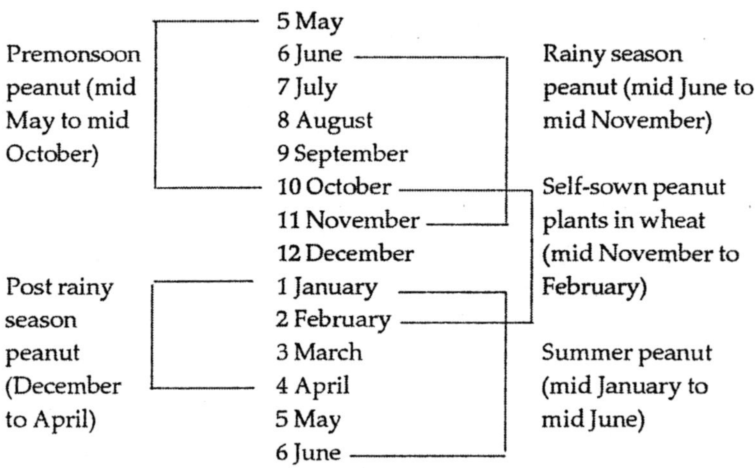


Fig. 1. Continuity of peanut crop helps rust pathogen to remain active throughout the year in Gujarat State.

Adequate degree of resistance against rust has been reported in genotype EC 76446 (292), NC Ac 17090, NC Ac 17132, PI 259747, PI 341879, PI 350680, PI 393646, PI 405132, PI 414322 etc. Certain of these genotypes possess resistance to late leaf spot disease as well (ICRISAT, 1987).

A successful control of rust by 0.2 to 0.3% mancozeb or chlorothalonil sprays is feasible (Dange *et al.*, 1981; Wangikar *et al.*, 1985). Systemic fungicides tridemorph (0.07%) and triademefon

(75 g/ha) control peanut rust (Mayee *et al.*, 1985). Against rust, nickle chloride too is equally effective (Seshadri, 1976). Rust and leaf spot diseases can be kept under check through sprays of carbendazim (0.05%) and mancozeb (0.2%) mixture (Dange *et al.*, 1981). Cheaper fungicides are required to be developed which can control both rust and leaf spot diseases.

## 4. Sclerotium Rot

In India, peanut Sclerotium rot is caused by *Sclerotium rolfsii* Sacc. The pod yield losses are around 27% (Chohan, 1974). When the crop is progressively maturing, brown lesions are visible on the stems at soil surface as well as below the soil line which later get covered by white cottony fungal growth impregnated by numerous mustard seed – like sclerotia. The leaves of infected plants turn yellow and assume characteristic upright position. The affected plants bear light weighted and shrivelled kernels which may get infected due to peg rotting.

Soil moisture, temperature and inoculum have a considerable effect on the disease development. The disease develops rapidly in sandy and sandy-loam, well aerated soils with 40 to 50% optimum water holding capacity (Dubey, 1958). Potassium fertilization minimizes the disease, whereas phosphorus maximizes it. Warm weather favours the disease development.

Plantavax and Vitavax control the disease effectively when drenched in sclerotia infested soil (Amma *et al.*, 1983) besides, Brassicol and Ceresan (Patil and Rane, 1982).

## 5. Dry Root Rot

*Corticium solani* (Prill. & Delacr.) Bourd & Galz. (*Rhizoctonia solani*) infection results into sunken brown lesions on primary roots. On stem near soil surface reddish-brown lesions enlarge and turn dark brown. Such a stem gives wire stem look (Feakin, 1973). Infected plants damped- off eventually. Damping-off of seedlings is not uncommon (Reddy and Rao, 1983).

The disease could be managed by seed dressing with carbendazim or carbendazium + thiram or captafol or thiram (Sharma and Bhowmik, 1982) or through mancozeb (Natrajan *et al.*, 1983).

Gypsum when applied to soil @ 150 kg/ha checked the reccurrence of the disease.

## MINOR FUNGAL DISEASES

### 1. Alternaria Leaf Spot and Leaf Blight

Particularly *Alternaria alternata* (Fr.) Keissler and *A. tenuissima* Kunze ex. Pers. Wilts. cause leaf spot and veinial necrosis on leaves amongst several *Alternaria* spp. (Vyas *et al.*, 1985).

### 2. Anthracnose

The anthracnose symptoms on leaves produced by *Colletotrichum dematium* (Pers. and Fries) Grov. Sensuvona appear as water- soaked yellowish spots which develop into typical blight symptoms upon severity of disease. Bordeaux mixture or any other copper oxychloride preparation effectively checks the disease (Singh *et al.*, 1975).

### 3. Pestaliopsis Leaf Spots

The pathogen *Pestaliopsis arachidis* Satya produces round brown spots on leaves (Satya, 1964).

### 4. Eye Spot

Eye spot disease is caused by *Emericella nidulans* (Eidam) Veill. (*Aspergillus nidulans*). The disease appears in the form of light brown typical eye shaped spots which frequently cover most of the lower part of leaf lamina (Bansal and Sobti, 1988).

Several other peanut leaf spot, blight producing and other pathogens are as under :

i) *Phyllosticta* sp. produces leaf spot in peanut, against which resitance has been reported in peanut cultivars (Ukey and Mehta, 1984).

ii) *Exerohilum halodes* (Drechsler) Leonard and Suggs anamorph of *Setosphaeria rostrata* Leonard causes leaf spot on summer season peanut crop and *Fusarium equiseti* (Corda) Sacc. in rainy season (Ghewande *et al.*, 1987).

iii) *Bipolaris specifera* (Bain) Subramm. (*Cochliobolus specifer* Nelson) causes leaf spots and blight on peanuts (Jha *et al.*, 1984).

iv) Phoma or Phomopsis blight. Two species of *Phoma* as *P.microspora, P. sorghina* (Sacc.) Boerema (Ghewande *et al.*, 1984) and *Phomopsis* sp. (Sharma, 1974) have been reported to cause leaf blight especially of the lower leaves. Mancozeb fungicide controls the disease effectively (Balasubramanian and Narayanasamy, 1983).

v) Fusariun wilt browning (necrosis) of vascular tissues of roots and stems is caused by *Fusarium oxysporum*. The tips of diseased plants typically droop. Wilted plants exhibit yellow drooped leaves (Feakin, 1973). Alongwith *F. oxysporum, F. solani* has also been found capable of causing damping-off disease (Reddy and Rao, 1980).

vi) Root discolouration : The roots of peanut plants infected by *Olpidium brassicae* (Wor.) Dang. remain stunted and show characteristic discolouration (Subramanyam and Mc Donald, 1980).

vii) Powdery mildew Of peanut *Oidium arachidis* has been reported for the first time in India by Ghewande and Reddy (1980).

viii) *Polymyxa graminis* Ledingham infects peanut roots when peanut plants are raised in soil infested with peanut clump virus (Nolt and Reddy, 1985).

## AFLATOXIN

In the United Kingdom alone in 1960, about 10,00,000 Turkey poults died due to unknown etiological disorder. This got dubbed Turkey 'X' disease. The toxic factor in the Brazilian groundnut meal infested with *Aspergillus flavus* has been extracted, isolated identified and named as aflatoxin (Vander Zijden *et al.*, 1962). All strains of *A. flavus* do not produce aflatoxin. Certain strains of *A. parasiticus* Spear also are shown to produce aflatoxin. Aflatoxins are potent carcinogenic, hepatotoxic, tetratogenic and immuno suppressive agents (Ciegler, 1975). Among all aflatoxins, $B_1$ is most toxic and commonly occurring one, than '$G_1$' '$B_2$' and '$G_2$' (Moss, 1972).

*A. flavus* gets an easy access in the kernels of peanut pods when harvested produce either gets wet due to unseasonal rains or when pods with high moisture content are stored. The problem aggravates further, if insect infestation and higher humidity occur during storage (Dange and Patel, 1984; ICRISAT, 1989). In 1979, India's contribution to the world net exports of peanut meal was about 64% which got down to 26.1% in 1988 due to imposition of strict tolerance limits of aflatoxin content by importing countries.

To overcome aflatoxin problem in peanuts and meal, the following methods are being adopted :

## 1. Prevention

The peanut crop should be harvested at the right stage and dried thoroughly with less than 8% moisture content before its storage (FAO, 1979). Avoid to store injured or insect damaged pods. Storage of peanut products at low temperature (< $10^{o}$ – $15^{o}$ C) and low RH (< 75%) also reduce the chance of fungal invasion to bare minimum. Another important approach is growing of resistant varieties, such as J-11, Karad 4-11, Ah 7233, Var. 27 (Dange and Prasad, 1989; Mehan *et al.*, 1986).

## 2. Removal

Electronic sorters, hand sorters and electronic eyes are under use these days in the U.S.A. to separate out discoloured and damaged kernels from healthy lots. Some chemicals like acetone and alcohol are employed to remove aflatoxin from peanut kernels and meal.

## 3. Detoxification

Ammonia, hydrogen peroxide and sodium hypochloride are certain potent chemicals employed to detoxify aflatoxin at a large scale (ICRISAT, 1989).

## BACTERIAL DISEASE

### Bacterial wilt :

*Pseudomonas solanacearum* E.F. Smith has been reported to cause bacterial wilt of peanut in India (Keshwal *et al.*, 1976; Khetmalas and Sardeshpande, 1989). Since the disease is of minor importance in India, much information is not available (Mehan *et al.*, 1985).

## VIRUS DISEASES

### 1. Bud Necrosis

Tomato spotted wilt virus (TSWV) of tomato causes a serious bud necrosis disease (BND) in peanuts in India (Iizuka and Reddy, 1986). It is widly distributed and causes heavy losses upto 50% and the incidence ranges from 5 to 100% (Rao *et al.*, 1979).

A wide variety of symptoms on peanut plants are produced by BND. Initially chlorotic rings appear on the young leaflets which later become necrotic. Necrosis of the terminal shoots and buds soon develops (chracteristic symptom) which may extend downwards killing the main axis. Though the virus can be transmitted by thrips *Frankliniella schultzei* Trybom and *Scirtothrips dorsalis* Hood, but *F. schultzei* is most effective vector (Reddy *et al.*, 1986). The virus is mechanically transmissible but it is not transmitted through seeds of groundnut in nature (Ghanekar *et al..*, *1979).*

Peanut genotypes NC Ac 1741, NC Ac 17888, NC Ac 343, C 108 and Robut 33-1 have exhibited field resistance to BND and show consistantly lower incidence of disease. High population of peanut plants and intercropping peanut with fast growing cereals such as pearl millets are means of decreasing BND incidence.

### 2. Clump

In sandy soils of Gujarat and Punjab states of India, the peanut clump disease was first observed in 1977. The disease is economically important in Punjab state. Early infected peanut plants fail to produce pods and in the event of late infection upto 60% losses are recorded (Ghanekar, 1980). The disease occurs in patches which extend year after year (being soil borne in nature) when peanut is planted in the same field.

Infected plants are severely stunted with a bushy look, bearing small dark green leaves. On young leaves chlorotic ring spots and mottling develop. Roots of infected plants become dark in colour. Of which the outer layers are peeled-off easily (Nolt *et al.*, 1988).

The virus is mechanically and graft transmissible. It has shown to be transmitted by a soil inhabiting fungus *Polymyxa graminis* Ledingh.

Soil application of carbofuran and Temik (aldicarb) prior to planting peanuts, significantly reduces disease incidence.

### 3. Mottle

The peanut mottle disease is caused by peanut mottle virus (PMV). The Virus has a narrow host range which infects mostly legumes. In peanut, upto 40% losses have been recorded. Young leaves characteristically show dark green islands interspersed with chlorotic lesions on *Phaseolus vulgaris* L. (Cv. Topcrop). It is transmissible through seeds upto 85% (Bharathan *et al.*, 1984). In nature, the mottle disease spreads through *Aphis craccivora* Koch and *Myzus persicae* Sulz.

### 4. Cowpea Mild Mottle

On peanut leaflets, chief symptoms of this disease appear as vein clearing with downward rolling and necrosis. The virus (CMMV) is trasmitted by white flies in nature (Iizuka *et al.*, 1984).

### 5. Yellow Spot

The disease is characterised by bright yellow spots on young leaflets which later turn necrotic. The virus is localized to leaves only. It is transmitted by thrips *Scirtothrips dorsalis* (Reddy *et al.*, 1986). Some times, the disease inflicts heavy losses. Details regarding control measures are lacking.

## 6. Chlorotic Leaf Streak and Yellow Mosaic

Peanut chlorotic leaf streak and peanut yellow mosaic disease are not of much economic importance (Moss and Reddy, 1989).

## 7. Green Mosaic

The disease is recognized by mosaic symptoms on leaves. Infected plants bear less pods with light kernels (Sreenivasulu *et al.*, 1981 ; Sreenivasulu and Nayadu, 1986). The virus (a potyvirus) is sap transmissible to different hosts. *Phaseolus vulgaris* is a good local lesion host. Seeds do not carry any virus (Sreenivasulu *et al.*, 1981).

## 8. Stripe

Peanut stripe virus causes striped green blotches on leaves Oak leaf mosaic symptoms have also been observed at many places in south India (Rao *et al.*, 1988).

## Mycoplasma Disease

Affected plants give a typical bushy look (Witches broom-like) from distance due to proliferation of axillary buds, smalling and stunting of leaves and transformation of floral parts into small leafy structures (Muthusamy and Subramanian, 1985).

The leaf hopper *Orosius argentatus* plays an important role in spreading witches' broom diseases in nature.

# NEMATODE DISEASE

## 1. Root Knot

*Meloidogyne arenaria* (Neal) Chitwood, *M. javanica* (Treub.) Chitwood and *M. incognita* (Kofoid and White) Chitwood are reported on peanut crop (Patel *et al.*, 1988). *M. arenaria* is rather more destructive (Chohan, 1978). The affected plants get stunted bearing yellow leaves with scorched margins.

The roots of such gently uprooted plants exhibit numerous corky knots of different sizes and shapes on pegs and pods. *M. arenaria* injuries provide sites on peanut pods for enhanced *Aspergillus flavus* infection. Second stage infective larvae survive in soil upto 18 months. A good control by seed treatment @ 3 to 6% seed weight basis with aldicarb, sulfone 75 WP and carbofuran 40F (flowable) culminates into reduced knotting, decreased final nematode population and improved plant growth (Patel *et al.*, 1986 a). Further, furrow application of phenamiphos (Nemacur) @ 2 kg a.i. / ha before sowing checks appreciably nematode population in soil. Peanut genotypes differ in their reaction to *M. arenaria* (Patel *et al.*, 1988).

## NUTRITIONAL DEFICIENCY

In certain areas, depending upon the type of soil, iron and sulphur deficiency symptoms have been observed.

### 1. Chlorosis

In Saurashtra region of Gujarat state, chlorosis due to iron deficiency is rather common in peanuts cultivated in medium black calcareous soils (Patel *et al.*, 1982). Characteristically green midribs and veinlets stand out against yellowish-white background of affected leaves. The growth of the plant is retarded and the leaves wither. A mixture of 1.0% ferrous sulphate and 0.1% citric acid, if sprayed 3 to 4 times at an interval of 12-15 days, the chlorosis is checked effectively. If soil is found deficient in iron an application of 50-100 kg/ha of ferrous sulphate makes up iron deficiency and gives a lushgreen look of the crop.

Spraying or soil application of ferrous sulphate alone is equally effective in protecting peanut crop from another form of chlorosis caused by iron linked sulphur deficiency.

## II SESAME (*Sesamum indicum* L.)

Sesame is affected by a number of diseases which cause considerable yield losses (Maiti *et al.*, 1985).

## FUNGAL DISEASES

### 1. Alternaria Leaf Spot

*Alternaria sesami* (Kawamura) Mohanty and Behera causes leaf spots on sesame is widely occurring in India (Deshpande and Shinde, 1976). The disease inflicts considerable yield losses.

Generally on leaves, conspicuous concentric rings develop as round to irregular dark brown spots Under favourable conditions, scattered spots coalesce to cover larger leaf area. The blighted leaves dry out and fall off prematurely.

*Alternaria alternata* causes characterstic bending of infected plants at lesions on stems and leaves which is an important feature (Rani *et al.*, 1984) of this disease.

In general, varieties differ in their reactions. Hairy varieties have been reported resistant (Dolle, 1984).

Besides, survival of pathogen in leaf debris buried in soil, the pathogen is seed borne too. The blight disease can be effectively checked by suitable seed treatment and spraying crop with 1% Bordeaux mixture, or 0.1% zineb (Samuel *et al.*, 1971).

### 2. White Leaf Spot

*Cercospora sesami* Zimmerman causing white leaf spot inflicts 5-20% yield losses in India (Brar and Ahuja, 1979).

Spots upto 5 mm in diameter with whitish centres and blackish- purple brown margins develop on both surface of leaves. When the disease is severe, the spots coalesce and premature defoliation occurs. On petioles too, elongated brown to black spots develop (Mohanty, 1958).

The pathogen survives through internally infected seeds and buried plant debris in soil (Leppik and Sowell, 1964). Hot water treatment of seeds at 53° C for 30 minutes eradicates internal infection to give a complete control of disease (Chowdhury, 1945). Carbendazim and Bayleton @ 1.5 g/kg seed are able to check the disease (Vyas, 1981).

### 3. Angular Leaf Spot

Mohanty (1958) made a first report of angular leaf spot due to *Cercospora sesamicola* Mohanty from Orissa state of India. Angular leaf spots measuring 1 to 8 mm in length and breadth get necrotized and become dark brown in colour. The pathogen is seed borne in nature (Ferrer, 1960). It also perpetuates as sclerotia in plant debris (Rathaiah and Pavgi, 1973).

## 4. Phytophthora Blight

Phytophthora bligh of sesame caused by *Phytophthora parasitica* (Dastur) var. *sesami* Prasad is responsible for heavy yield losses in many states of India. As high as 72 to 79% plant mortality has been reported due to this disease only (Gemawat and Prasad, 1964).

Deep brown to black enlarged spots appear on leaves and stems. When conditions are humid white cottony mycelial growth is visible over infected capsules. The capsules bear shrivelled and light weighted seeds. If infection occurs at the seedling stage, the seedlings die.

The fungus apparently survives as chlamydospores on infested seeds (Sehgal and Prasad, 1966). Late planted crop is less prone to disease (Rathaiah, 1983).

Sources of resistance in certain plant types have been identified (Sehgal and Prasad, 1972). Seed treatment with thiram (0.3%) eliminates seed infection to a great extent. Spraying of Bordeaux mixture, or mancozeb, or zineb, or copper based fungicides are reported effective against blight disease (Vyas, 1981).

## 5. Powdery Mildew

Various powdery mildew fungal pathogens like *Oidium erysiphoides* Fr. (Roy, 1965); *O. acanthospermi* (Krishnaswami *et al.*,. 1984), *Sphaerotheca fuliginea* (Schlecht) Pollacci (Gemawat and Verma, 1972) and *Leveillula taurica* (Lev.) Arnaud (Patel *et al.*, 1949) on sesame have been reported from India. Generally, dirty white, superficial and powdery fungal growth is confined to upper surface of infected leaves.

Sesame genotypes differ in their reactions. A few resistant genotypes have been identified (Gaikwad *et al.*, 1977). The crop can be saved by dusting 20 kg/ha sulphur dust or by spraying 0.2% wettable sulphur (Shanmugam *et al.*, 1976).

## 6. Fusarium Wilt

*Fusarium oxysporum* (Schelt) *f. sp. sesami* Jacq. is able to infect plants at any growth stage. The leaves of infected plants turn yellowish and show epinasty. The affected plants finally wither.

Freshly affected plants show internal discolouration of vascular tissues. The root rots at a later stage.

The pathogen is able to survive for many years in soil, besides being seed borne in nature. To ensure disease free crop the seed material must invariably be given treatment with thiram @ 2-3 g/kg seeds.

### 7. Charcoal Rot

*Macrophomina phaseolina* (Tassi) Goid causes charcoal rot disease in sesame. Another pathogenic isolate *M. phaseolina* f.sp. *sesamica* (Kulkarni *et al.*, 1966) also infects sesame plants in fields. Any stage of the crop is vulnerable to infection. Early infection at the seedling stage results into death of seedlings. Late infection causes abundant sclerotia and pycnidia formation at the blackened collar regions of infected plants. The leaves droop. Affected plants remain stunted or may topple down. Infected capsules open prematurely exposing light weighted discoloured and shrivelled seeds (Mishra *et al.*, 1973). The fungus is carried over through soil or seeds (Kushi and Khare, 1978). The pathogen possesses a wide host range.

A complete control of seed borne *M. phaseolina* infection is possible through seed treatment either with captafol, or thiram or carbendazim @ 3 g/kg seed (Virk and Grover, 1975). Since the pathogen is a soil inhabitant, a long crop rotation with nonhost plants and summer ploughing are helpful to reduce the inoculum. Black and green seeded varieties are less susceptible as compared to white seeded varieties (Bartaria, 1988).

## BACTERIAL DISEASES

### 1. Bacterial Blight

*Xanthomonas campestris* (Pammel) Dowson pv. *sesami* (Sabet and Dowson) Dye inflicts heavy yield losses in sesame (Chand *et al.*, 1970).

Seedling mortality occurs at the seedling stage (Rao, 1962). On leaves numerous scattered water soaked spots coalesce, necrotize and turn dark brown to black in colour. The blighted leaves become brittle, dry out and finally defoliate. The disease is local but may become systemic when infection occurs through petioles and stems resulting into death of plants (Rao and Durgapal, 1966).

The pathogen is an internally seed borne (Rao and Durgapal, 1966). It has been found to be surviving on weed *Acanthospermum hispidum* DC (Nayak and Sharma, 1981). Cotyledonary seedling mortality is common (Rao, 1962). Leaves exhibiting necrotic, dark brown to black coloured spots develop typical blight symptoms. Such leaves become brittle dry out and finally defoliate. When disease is severe it appears also on stems and capsules. Infected stems turn brown. The death of plants occur (Rao and Durgapal, 1966).

## 2. Bacterial Leaf Spot

Bacterial leaf spot is prevalent in most part of the sesame growing tracts in India (Singh, 1970). It is caused by *Pseudomonas syringae* van Hall pv. *sesami* (Malkoff) Young, Dye and Wilkie. The disease has been reported to cause 21 to 27% losses in grain yield under simulated field conditions (Vajavat and Chakravarti, 1978).

Under high rainfall and humidity conditions, necrotic purplish to brownish black coloured spots appear on stems, petiole bases and capsules, besides, leaves where the spots advance along the veins and petioles.

The pathogen is internally seed borne (Durgapal *et al.*, 1869). It remains viable in seeds for 11 months. In soil the pathogen may survive for 1-3 months only. The best control of disease is obtained by 100 ppm Agrimycin seed treatment followed by a spray of 100 ppm Agrimycin when the disease appears first in the field (Verma and Daftri, 1976).

## 3. Bacterial Wilt

Bacterial wilt (*Pseudomonas solanacearum*) has also been observed in natural conditions (Ansari and Ram, 1987).

## VIRUS DISEASES

### 1. Leaf Curl

In most of the sesame growing areas of India, the leaf curl virus disease causes heavy losses and reduces the oil content of seeds appreciably (Prasad *et al.*, 1985). The disease is caused by Nicotiana virus 10 on leaves. Certain characteristic symptoms of typical leaf curl disease appear as vein thickening, puckering, curling, reduction in laminar size and leatheriness. The causal virus also infects papaya, tobacco and tomato in fields. It is transmitted by white fly (*Bemisia tabaci* Genn.) and grafting. Certain resistant genotypes have been identified (Jani and Bharodia, 1978).

## MYCOPLASMA DISEASE

### Phyllody

Mycoplasma-like organisms (MLO's) associated with sesame phyllody (Purohit *et al.*, 1978) have been reported from many parts of India (Vasudeva and Sahambi, 1955). Higher yield losses due to this disease in India are 39 to 74% (Brar and Ahuja, 1979). The characteristic symptom is the transformation of floral parts into typical green leafy sterile structures resulting into green fascicles.

The pathogen is transmitted by leaf hopper *Orosius albicinctus* Distant which also infects chickpea, rapeseed, etc. in fields (Bindra and Singh, 1970). The pathogen survives in these hosts which act as reservoirs for insects.

Tetracycline antibiotic sprays (500 ppm) give temporary recovary from disease. Thimet 10 G, or phorate 10 G, or temik 10G applied at sowing time in soil combined with spraying of 0.1% metasystox proved very effective control measure (Muheet and Chauhan, 1975).

## III SUNFLOWER (*Helianthus annus* L.)

In India sunflower (*Helianthus annus* L.) cultivation is progressively increasing. The crop is attacked by several diseases which result into heavy yield losses.

## FUNGAL DISEASES

### 1. Seed Rot and Seedling Blight

Several pre- and post-emergence deaths of sunflower seedlings reflected in the form of damping-off, root rot, stem rot diseases in the fields result into poor and patchy plant population. To this sort of menace *Aspergillus niger, Pythium aphanidermatum, Rhizoctonia, Rhizopus stolonifer, R. arrhizus, Sclerotium rolfsii* are found associated.

Thiram or mancozeb or captafol @ 3 g/kg seed treatment protects effectively sunflower crop against seed rot and seedling diseases (Kolte, 1974).

### 2. Alternaria Blight

Besides *Alternaria helianthicola* Rao and Raj (Rao and Rajgopalan, 1977) and *A. tenuissima* (Kunzeex. Pers.) Wiltshire (Narain and Chauhan, 1981) *Alternaria helianthi* (Hansf.) Tubaki and Nishihara has been found more destructive to sunflower crop in India (Mukewar *et al.*, 1974). The seed yield loss is 27 to 80% in India (Balasubrahmanyam and Kolte, 1980).

Firstly on lower leaves, circular to oval, dark brown to black spots of 0.2 to 5 mm in diameter marked with concentric rings surrounded by chlorotic zones develop which spread progressively under humid conditions to cause blighting of leaves, stems and flower heads.

The pathogen obviously survives in plant debris and seeds (Raut, 1985 a). Spraying of 0.3 mancozeb has proved quite effective (Kolte *et al.*, 1979) over zineb, ziram, captafol, miltox, etc.

### 3. Brown Leaf Spot and Blight

Saharan and Singh (1976) have reported brown leaf spot and blight disease from India. The pathogen *Septoria helianthi* Ellis and Kelleman initially produces small, irregular, yellow to brown spots of about 10-15 mm in diameter. Under favourable conditions, necrotic spots engulfed by lighter periphery coalesce to form irregular blotches. Such leaves get blighted and defoliate. The pathogen invariably survives in host debris buried in soil. The perpetuation of pathogen in seeds cannot be ruled out.

### 4. Downy Mildew

*Plasmopara halstedii* (Farl.) Berl. and de Toni is the cause of downy mildew disease of sunflower. Though the disease is confined to the Maharashtra state of India, it inflicts heavy yield losses in other countries of the world.

Depending upon the stage of crop, the infection may result in damping-off, local lesions on leaves and basal root, or stem galls and even systemic disease. Systemically infected plants remian stunted with chlorotic leaves. Under humid conditions the downy growth comprised of sporangiophores and sporangia is conspicuous on the lower surface of leaves. Rarely the flowers form on infected heads. If formed, they remain sterile or bear no seeds. Under severe conditions of disease, the roots get discoloured and hypertrophied. Melero-Vara *et al.* (1982) have shown the efficacy of metalaxyl seed treatment @ 3-6 g/kg seed. If the seed treatment is followed by spray of metalaxyl the disease can be managed easily. A formulated product Phytoalexin when used as spray, helps reducing seedling mortality and downy mildew incidence (Mulekar *et al.*, 1988).

Since the disease is confined to Maharashtra, strict quarantine measures can check its introduction into other sunflower growing areas of the country.

### 5. Rust.

The pathogen *Puccinia helianthi* Schw. infects lower leaves of sunflower underneath of which orange-black uredopustules appear. Under severe conditions, the younger leaves, stems, petioles, floral bracts and flowers are also infected. In India, all the spore forms (pycnial and aecial stages are not common) are produced on annual and perennial *Helianthus* species to complete the life cycle of pathogen (Mathar *et al.*, 1975). Black coloured teliopustules develop late in the season.

The variety EC 82819 proved resistant against this disease (Mathur *et al.*, 1982).

Effective control of rust is possible by spraying 0.1 to 0.2% mancozeb or maneb or zineb (Bhowmik and Singh, 1979). Systemic fungicides benodanil and oxycarboxin have also been found effective (Mathur *et al.*, 1984).

## 6. Powdery Mildew

The disease is mainly caused by *Erysiphe cichoracearum* DC f. sp. *helianthi* Jacz. and by another fungus *Sphaerotheca fuliginea* (Schlecht. ex. Fr.) Pollacci in certain areas (Raut and Kale, 1976).

A large area of mainly upper and lower surface of leaves is covered by whitish powdery growth under moderate conditions of disease development. Affected leaves curl, turn chlorotic-brown and finally collapse. The cleistothecia, if formed appear as minute black fruiting bodies on the lower surface of moribund leaves through which the pathogen apparently survives in plant debris or on perennial species of *Helianthus.*

A hyperparasite *Cicinnobolus cesatii* de Bary is observed to parasitize *S. fuliginea* (Chona and Munjal, 1956) which obliterates cleistothecial formation.

## 7. Charcoal Rot

*Macrophomina phaseolina* (Tassi) Goid affected sunflower plants get wilted. The estimates of loss due to the disease are lacking (Sivaprakasam *et al.*, 1975). The pathogen also causes seedling blight, damping-off, root rot, and basal stem rot (Fakir *et al.*, 1976) when the soil is kept wet. The pathogen survives through sclerotia which are either carried through seeds or buried in soil in plant debris (Raut, 1985).

Seed treatment with thiram and benomyl @ 3-4 g/kg seed very effectively checks the seed borne infection (Raut and Bhombe, 1986). Spray chemicals have not been tested as yet.

## 8. Rhizopus Head Rot

The disease caused by *Rhizopus arrizus* Fisher (Agrawat *et al.*, 1978) takes a heavy toll in wet weather. Injury to flower heads is essential for initial infection to establish (Mishra *et al.*, 1972). The larvae of *Heliothis armigera* cause injury to sunflower heads for pathogen infection (Agrawat *et al.*, 1978).

## 9. Sclerotinia Root Rot or Wilt

Various sunflower diseases like root rot or wilt, stem rot or stalk rot and head rot are caused by *Sclerotinia sclerotiorum* (Lib.) de Bary. Typical symptoms as root rotting occur at the time of seed setting after the appearance of water soaked lesions on tap root and later on stem near soil surface. On the infected tissues, white fungal growth impregnated with scattered black coloured sclerotia is conspicuous. Finally the leaves turn yellow and the plant wilts (Dorrell and Huang, 1978).

The sclerotia survive in soil and on contaminated seeds. At flowering time the ascospores produced after germination of sclerotia are responsible for air borne infection to cause stalk or head rots in sunflower plants (Huang and Dueck, 1980).

Carbendazim or thiophanate methyl or benomyl or thiram or mancozeb or zineb seed treatment effectively checks seed or soil borne infection (Sharifi-Tehrani, 1974). Air borne infection is manageable by spraying mancozeb or thiophanate methyl or benomyl or dichloran.

## MINOR FUNGAL DISEASES

1. *Botrytis cinerea* Pers. ex. Fr. causes collar rot in India. (Mathur *et al.*, 1981).
2. *Cladosporium cladosporoides* (Fres.) De Vries is responsible for leaf spot (Anilkumar and Seshadri, 1975).
3. Drechslera leaf spot disease occurs due to *Drechslera hawaiiensis* (Bugnicourt) Subram. and Jain ex. M.B. Ellis (Reddy and Siradhana, 1978).
4. The leaf blight disease is caused by many pathogens viz.
   - i) *Botryodiplodia theobromae* Pat. (Saksena *et al.*, 1976),
   - ii) *Thanatephorus cucumeris* (Frank) Dark (Chauhan, 1980),
   - iii) *Phoma helianthi* (Pachkhede *et al.*, 1985),
   - iv) *Fusarium equiseti* (Corda) Sacc. and *F. solani* respectively, cause root rot and collar rot diseases in India (Gupta *et al.*,

1980).

5. *Myrothecium roridum* Tode ex. Fr. causing leaf spot disease, has been reported for the first time on sunflower in India (Narain and Koul, 1984).

## BACTERIAL DISEASES

**1. Bacterial leaf spot**

*Xanthomonas campestris* pv. *phaseoli* (Smith) Dye causes leaf spotting on sunflower (Kore and Biradar, 1986). SC 17 variety proved highly resistant.

## VIRUS DISEASES

**1. Sunflower mosaic**

Two distinct sunflower mosaic diseases have been reported.

a) The mosaic is associated with ring spots. The virus is transmitted by sap and infects sunflower crop only (Battu and Phatak, 1965).

b) The mosaic symptoms are associated with malformation and cupping of leaves. The virus is transmitted through seed and white fly *Bemisia tabaci* Genn. besides, sap and graft. The virus is capable to infect *Amaranthus* sp, systemically. *Chenopodium amaranticolor* is a local lesion host (Gupta, 1981).

**2. Yellow ring mosaic**

Affected plants remain stunted, bear malformed young leaves and less number of flowers. The virus is sap and graft transmitted (Gupta, 1981).

**3. Yellow spot**

On leaves characteristic yellow spots are produced. The virus is transmitted by grafting (Gupta and Gupta, 1977).

## PHANEROGAMIC PARASITE

*Orobanche cernua* Loeff. var. *desertorum* Black. is common on sunflower roots where it causes pinhead-sized infection (Krishnamurty *et al.*, 1977).

## NEMATODE DISEASES

Roots of sunflower plants are found attacked by *Rotylenchulus reniformis* and *Meloidogyne incognita* nematodes (Yadav and Pathak, 1985).

## IV NIGER (*Guizotia abyssinica* Cass.)

In India, major Niger (*Guizotia abyssinica* Cass.) producing states are Andhra pradesh, Madhya pradesh, Maharashtra and Orissa. Seemingly the crop is affected by many diseases. Owing to its limited production, the information on various diseases are scanty.

## FUNGAL DISEASES

### 1. Collar rot disease

Affected wilted seedlings show sunken collar rotting which is generally covered by white mycelial mat impregnated by abundant chocolate sclerotial bodies. The disease is caused by *Corticium rolfsii* Curzi. (Siddaramaiah *et al.*, 1979).

### 2. Cercospora leaf spot

Govindu and Thirumalachar reported Cercospora leaf spot of niger in 1956. The disease is caused by *Cercospora guizoticola* Govindu and Thirumalachar. Small straw to brown coloured spots appear on leaves which coalesce under favourable weather conditions.

The pathogen perpetuates through seed and plant debris buried in soil. Seed borne infection can effectively be prevented by treating seeds with thiram @ 3 g/kg seed. Carbendazim spray @ 0.025% checks the disease.

### 3. Curvularia leaf spot

In India, Kore and Shirurkar (1984) have reported *Curvularia lunata* causing leaf spot disease on niger crop.

### 4. Alternaria leaf spot.

*Alternaria* sp. causing leaf spots on niger leaves can be effectively managed by 0.2% zineb sprays (Sharma, 1982).

**5. Powdery mildew**

Fungus *Sphaerotheca* forms bigger irregular cottony spots on leaves. Eventually, affected leaves defoliate (Vyas, 1981). The pathogen survives on collateral hosts. Wettable sulphur fungicides when sprayed @ 0.2 to 0.3% effectively control this disease.

**6. Ozonium wilt**

Typical symptoms of Ozonium wilt are observed as necrotic patches near soil surface on the stem of older plants. The disease is caused by *Ozonium texanum* Neal and Wester var. *parasiticum*. Under humid conditions, the mycelial growth develops on necrotic patches. The stem may break at the affected region (Rao and Pavgi, 1975). When the disease becomes severe, the twigs and inflorescences start rotting.

**7. Root rot**

*Macrophomina phaseolina* (Maubl.) Ashby is responsible to cause root rot disease in niger seedlings in India (Sharma, 1982).

**8. Sclerotium rot**

Besides *Sclerotium rolfsii* Sacc. (Agrawat, 1979) causing root and stem rotting of niger, other fungi such as *Rhizoctonia bataticola* (Taub.) Butler, *Alternaria spp., Aspergillus flavus* (Desai *et al.*, 1981, Sharma, 1982) have been found associated with seed rot and seedling blight.

## BACTERIAL DISEASE

### Bacterial leaf spot

*Xanthomonas campestris* pv. *guizotiae* var. *indicus* Moniz, Syed and Raj. causes brown spots on leaves which are surrounded by yellow halos. These spots later turn out as irregular blotches (Moniz *et al.*, 1968).

## PHANEROGAMIC PARASITE

A phanerogamic parasite *Cuscuta hyalina* causes severe stunting of niger plants. Such plants turn pale- yellow and bear less number of flowers and fruits. Since the parasite gets access in the field through seeds, a careful separation of Cuscuta seeds from seed material and removal and destruction of the parasite before it flowers in the field is a must for its eradication (Vyas, 1981).

In the Orissa state of India *C. chinensis* Lamk causes heavy losses in seed yield (Mohanty and Rath, 1987) of niger.

## REFERENCES


Agrawat, J.M. (1979) Sclerotium rot of *Glycrrhiza glabra* – a new host record for Rajasthan. Sci. Cult. 45 : 366.

Agrawat, J.M., Vaish, O.P., Mathur, S.J. and Chippa, H.P. (1978) Some observations on Rhizopus head rot of sunflowers in Rajasthan, India. 8th International. *Sunflower Conf.*, Minneapolis, Minn., USA, 23 pp. (Abstr.).

Amma, P.A., Shanmugam, N. and Nair, M.C. (1983) Studies on the control of sclerotial root rot of groundnut. *In* proceedings of the National Seminar on Management of disease of oilseed crops, Madurai, India, 44-45 pp.

Anil Kumar, T.B. and Seshadri, V.V. (1975) Cladosporium leaf spot of sunflower. Curr. Sci. 44: 722.

Ansari, M.M. and Ram, T. (1987) Bacterial wilt of sesamum caused by *Pseudomonas solanacearum*, a new record for Andaman and Nicobar Island. Indian Phytopath. 40 : 236.

Balasubrahmanyam, N. and Kolte, S.J. (1980) Effect of Alternaria blight on yield components, oil content and seed quality of sunflower. Indian J. agric. Sci. 50: 701-706.

Balasubrahmanian, P. and Narayanasamy, P. (1983) Evaluation of fungicides for their efficiency against blight diseases of groundnut. *In* proceedings of the National Seminar on Diseases of Oilseed crops of Madurai, India, 41-43 pp.

Bansal, R.K. and Sobti, A.K. (1988) Eye spot diseases of groundnut. Indian J. Mycol. Plant Pathol. 18 : 218.

Bartaria, A.M. (1988) Association of seed coat with resistance to *Macrophomina phaseolina* in sesamum cultivar. J Indian Bot. Soc. 67 : 222.

Battu, A.N. and Phatak, H.C. (1965) Observations on a mosaic diseases of sunflower. Indian Phytopath. 18 : 317.

Bharathan, N., Reddy, D.V.R., Rajeshwari, R., Murthy, V.K. and Rao, V.R.

(1984) Screening peanut germplasm lines by enzyme linked immunosorbent assay for seed transmission of peanut mottle virus. Plant Dis. 68 : 757-758.

Bhowmik, T.P. and Singh, Amar. (1979) Evaluation of certain fungitoxicants for the control of sunflower rust. Indian phytopath. 32 : 443-444.

Bindra, O.S. and Singh, J. (1970) Biology and bionomics of *Orosius albicinctus* Distant, the jassid vector of sesamum phyllody virus. Indian J. agric. Sci. 48: 340-355.

Brar, G.S. and Ahuja, K.C. (1979) Sesame : Its culture, genetics, breeding and biochemistry. *In* Annu. Rev. Plant Sci. Vol.1, Malik, C.P. Edn., Kalyani Publishers New Delhi, India, 245 pp.

Chahal, A.S. and Chohan, J.S. (1971) Puccinia rust on groundnut. FAO plant prot. Bull. 19 : 90.

Chand, J.N., Mandloe, S.C. and Kulkarni, S.N. (1970) Bacterial leaf blight of Sesamum in Madhya Pradesh. Indian Phytopath. 23 : 156.

Chauhan, L.S. (1980) Rhizoctonia leaf blight of sunflower. Indian Phytopath. 33 : 606.

Chohan, J.S. (1974) Recent advances in diseases of groundnut in India. 171-184 pp. (Eds. Raychaudhuri S.P. and Verma J.P.), Current trends in plant pathology, S.N. Das & Gupta Birthday celebrations committee, Lucknow, 341 pp.

Chohan, J.S. (1978) Diseases of oil seed crops, future plans and strategy for control under small holdings. Indian Phytopath. 31 : 1-15.

Chona, B.L. and Munjal, R.L. (1956) *Cicinobolus cesatii* De Bary. hyperparasite of powdery mildews in India. Indian phytopath. 9: 98-105.

Chowdhury, S. (1945) Control of Cercospora blight of til. Indian J. agric. Sci. 15 : 140-142.

Ciegler, A. (1975) Mycotoxins : Occurrence, chemistry, biological activity. Lloydia 38 : 21-34.

Dange, S.R.S. and Desai, N.D. (1985) Screening of groundnut germplasm for rust resistance and search for the rust pathogen teliospores in Gujarat state. Agric. Sci. Digest 5 : 227-228.

Dange, S.R.S., Desai, N.D. and Patel, J.R. (1981) Chemical control of groundnut leaf spots and rust in Gujarat. *Indian Phytopath.* 34 : 110 pp. (Abstr.).

Dange, S.R.S. and Patel, V.J. (1984) Effect of relative humidity and storage period on fungal invasion and viability of groundnut seeds. Bull. Grain Technol. 22: 225-231.

Dange, S.R.S. and Prasad, S.R. (1989) Aflatoxin production on seeds of selected groundnut cultivars. Bull. Grain Technol. 27: 60-62.

Dange, S.R.S. and Saradava, M.R. (1985) Screening of selected groundnut cultivars against seed rot and collar rot caused by *Aspergillus niger*. Agric. Sci. Dig. 5 : 138-140.

Dange, S.R.S. and Saradava, M.R. (1986) Effect of plant age on collar rot of groundnut caused by *Aspergillus niger*. Indian J. Mycol. Plant Pathol. 16: 287-288.

Dange, S.R.S. and Saradava, M.R. (1988) Effect of soil types and soil moisture levels on seed rot and collar rot caused by *Aspergillus niger* in groundnut. Indian Phytopathol. 20 : 285 pp (Abstr.).

Desai, S.A., Siddaramaiah, A.L. and Hegde, R.K. (1981) Effect of culture filtrate of five species of fungi on seed germination and root and shoot elongation of niger. Mysore J. Agric. Sci. 15 : 48-50.

Deshpande, G.D. and Shinde, D.D. (1976) Occurrence of cultural strains of *Alternaria sesami* (Kawamura) Mohanty and Behera in Maharastra. J. Maharastra Agric. Univ. 1: 124.

Dolle, U.V. (1984) Studies on leaf blight of sesame (*Sesamum indicum* Linn.) caused by *Alternaria sesami* (Kawamura) Mohanty and Behera. Mysore J. Agric. Sci. 18 : 89-90.

Dorrell, D.G. and Huang, H.C. (1978) Influence of Sclerotinia wilt on seed yield and quality of sunflower wilted at different stages of development. Crop Sci. 18 : 974-976.

Dubey, H.D. (1958) Relation of soil texture and occurrence of root rot disease (*Sclerotium rolfsii* Sacc.) of peanut. Plant Dis. Reptr. 42 : 1376-1377.

Durgapal, J.C., Patel, P.N. and Rao, Y.P. (1969) Resistance in crop to bacterial disease in India I. Evaluation of sesamum for resistance to bacterial leaf spot disease incited by *Pseudomonas sesami*. Indian Phytopath. 22 : 292-294.

Fakir, G.A., Rao, M.H. and Thirumalachar, M.J. (1976) Seed transmission of *Macrophomina phaseolina* in sunflower. Plant Dis. Reptr. 60 : 736.

FAO (1979) Perspective on Mycotoxins. FAO Food and Nutrition Paper

No.13, 167 pp.

Feakin, S.D. (Ed.) (1973) Pest control in groundnuts. PANS Manual No.2 C O P R, Foreign and Commonwealth office, Overseas Development Admin., London, 197 pp.

Ferrer, J.B. (1960) The occurrence of angular leaf spot of sesamum in Panama. Plant Dis. Reptr. 44 : 221.

Gaikwad, S.J., Sabley J.E. and Buldeo, A.N. (1977) Reaction of some sesame (*Sesamum indicum* L.) varieties to powdery mildew disease at Nagpur. J. Maharastra Agric. Univ. 2 : 170-171.

Gemawat, P.D. and Prasad, N. (1964) Further studies on Phytophthora blight of sesamum. Indian Phytopath. 17 : 273-283.

Gemawat, P.D. and Verma, O.P. (1972) A new powdery mildew of *Sesamum indicum* incited by *Sphaerotheca fuliginea*. Indian J. Mycol. Plant Pathol. 2 : 94.

Ghanekar, A.M. (1980) Groundnut virus research at ICRISAT. *In* Proceedings of International workshop on groundnut. (Gibbons, R.W. and Mertin, J.V. Eds.) International crops Research Institute for the Semi-arid Tropics (ICRISAT), India, 211-216 pp.

Ghanekar, A.M., Reddy, D.V.R., Iizuka, N., Amin, P.W. and Gibbons, R.W. (1979) Bud necrosis of groundnut (*Arachis hypogaea*) in India caused by tomato spotted wilt virus. Anal. Appl. Biol. 93 : 173-179.

Ghewande, M.P., Jhala, Raj., Pande R.N. and Shukla, A.K. (1987) Leaf spots of groundnut. Indian Phytopath. 40 : 432-433.

Ghewande, M.P., Pandey, R.N., Shukla, A.K. and Misra, D.P. (1982) Spore germinability of *Aspergillus* species in seed coat leachates of groundnuts. Ind. J. Mycol. Plant Pathol. 14 : 134-137.

Ghewande, M.P., Pandey, R.N., Shukla, A.K. and Misra D.P. (1984) Toxicity effect of culture filterates of *Aspergillus flavus* on germination of seed and seedling growth of groundnut. Indian Bot. Reptr. 3 : 107-111.

Ghewande, M.P. and Reddy, P.S. (1987) Powdery mildew-a new disease of groundnut in India. Curr. Sci. 56 : 196.

Ghewande, M.P., Shukla, A.K., Pandey, R.N. and Misra, D.P. (1983) Losses in groundnut yield due to leaf spots and rust at different intensity level. Indian J. Mycol. Plant Pathol. 15 : 125-127.

Ghuge, S.S., Mayee, C.D. and Godbole, G.M. (1981) Assessment of losses in peanut due to rust and tikka leaf spots. Indian Phytopath. 34 : 179-182.

Godara, S.L. and Thakore, B.B.L. (1987) Control of mixed infection of leaf spots and rust diseases of groundnut by triadimefon and triadimenol, their absorption, translocation and persistence. Summa Phytopathologica 13 : 122-132.

Gupta, K.C. (1981) Studies on the identity of sunflower mosaic virus. III Intern. Symp. Plant Pathol., New Delhi, 117 pp.

Gupta, K.C. and Gupta, M.N. (1977) Viral diseases of sunflower (*Helianthus annus* L.), New Records from India. Curr. Sci. 46 : 285.

Gupta, K.S., Roy A.N. and Gupta, M.N. (1980) *Fusarium solani* incited collar rot of sunflower (*Helianthus annus* L.) - a new record. Sci. Cult. 46: 144.

Hennen, J.F., Figueiredo, M.B., Riberiro, I.J.A. and Soave, J. (1976) The occurrence of teliospores of *Puccinia arachidis* (Uredinales) on *Arachis hypogaea* in Sao Paulo State, Brazil. Summa Phytopathologica 2 : 44 -46.

Huang, H.C. and Dueck, J. (1980) Wilt of sunflower infection by mycelial germinating sclerotia of *Sclerotinia sclerotiorum*. Can. J. Plant Pathol. 2 : 47-52.

ICRISAT (International Crops Research Institute for the Semi-arid Tropics) (1987) Groundnut rust disease. Proceedings of a Discussion Group Meeting (24-28 Sept., 1984), Patancheru, A.P., India, 204 pp.

ICRISAT (International Crops Research Institute for the Semi-Arid Tropics) (1989) Aflatoxin contamination of groundnut : Proceedings of the International Workshop (6-7 Oct., 1987), Patancheru, A.P., India 423 pp.

Iizuka, N., Rajeshwari, R., Reddy, D.V.R. Goto, T., Munizappa, V., Bharathan, N. and Ghanekar, A.M. (1984) Natural occurrence of strain of cowpea mild mottle virus on groundnut (*Arachis hypogaea*) in India. Phytopath. Z. 109 : 245-253.

Iizuka, N. and Reddy, D.V.R. (1986) Identification of viruses from peanut in India. Tech. Bull. Trop. Agric. Res. Cen. No. 21: 164-183.

Jani, S.M. and Bharodia, R.K. (1978) Varietal susceptibility of sesamum (*Sesamum indicum*) to leaf curl disease. Gujarat Agric. Univ. Res. J. 4 : 28-29.

Jha, D.K., Dutta, A.K. and Rai, K.K. (1984) A new leaf spot disease of groundnut. Trop. Grain Legume Bull. No. 29 : 37-38.

Kenjale, L.D., Shinde, P.A. and Utikar, P.G. (1981) Loss in groundnut yield due to rust. Indian Phytopath. 38 : 374-377.

Keshwal, R.L., Kulkarni, S.N. and Jain, A.C. (1976) Occurrence of new strain of *Pseudomonas solanacearum* E. F. Smith in Madhya Pradesh infecting groundnut (*Arachis hypogea* L.) and Chilli (*Capsicum annum* L.) J.N.K.V.V. Res. J. 10 : 89.

Khetmalas, M.B. and Sardeshpande, J.S. (1989) Bacterial wilt of groundnut in Maharashtra. J Maharashtra Agric. University 14 : 369.

Kolte, S.J. (1974) Influence of fungicidal seed treatment on emergence of sunflower under field conditions. Pesticides 8 : 45-46.

Kolte, S.J. (1985) Diseases of Annual oilseed Crops. Vol. I, Peanut diseases, 143 pp; Vol. II, Rapseed - Mustard and sesame diseases, 135 pp; Vol. III, sunflower, Safflower and Nigerseed diseases, 154 pp., C.R.C. Press, Inc. Boca Raton, Florida.

Kolte, S.J., Balasubramanyam, N., Tewari, A.N. and Awasthi, R.P. (1979) Field performance of fungicides in the control of Alternaria blight of sunflower. Indian J. Agric. Sci. 49 : 555- 559.

Kore, S.S. and Biradar, M.V. (1986) Bacterial leaf spot diseases of sunflower. Maharashtra Agric. Univ. 11 : 187-189.

Kore, S.S. and Shirurkar, P.D. (1984) *Curvularia lunata* on niger -new record. Indian J. Mycol. Plant Pathol. 14 : 193.

Krishnamurty, G.V.G., Lal, R. and Nagrajan, K. (1977) Further studies on the effect of various crops on the germination of *Orobanche* seed. PANS 23 : 206-208.

Krishnaswami, S., Rao, A.V. and Appadurai, R. (1984) Resistance to powdery mildew in sesamum. Indian. J. agric. Sci. 54 : 666- 668.

Kulkarni, N.B., and Patil, B.C. (1966) Taxonomy and discussion on the nomenclature of *Macrophomina phaseoli* (Maubl). Ashby, and its isolate from India. Mycopath. Mycol. Appl. 28 : 257-264.

Kushi, K.K. and Khare, M.N. (1978) Comparative efficacy of five methods to detect *Macrophomina phaseolina* associated with sesamum seeds. Indian Phytopath. 31 : 258-259.

Leppik, E.E. and Sowell, G. (1964) *Alternaria sesami*, a serious seed-brone pathogen of world-wide distribution. FAO plant Prot. Bull. 12 : 13-16.

Lewin, H.D. and Natrajan, S. (1971) Fungicidal control of tikka leaf spot of groundnut in Tamil Nadu. Madras Agric. J. 58 : 480- 487.

Mathar, A.S., Subrahmanyam, C.L. and Ramasamy, R. (1975) Occurrence of aecial and pycnial stages of *Puccinia helianthi* Schw. in India. Curr. Sci. 44 : 284.

Maiti, S., Raoof, M.A., Sastry, K.S. and Yadava, T.P. (1985) A review of sesamum diseases in India. Trop. Pest. Management. 31 : 317-323.

Mathur, S., Thakore, B.B.L., Chakravarti, B.P. and Singh, R.B. (1981) Outbreak of collar rot of sunflower caused by *Botrytis cinerea* in Rajasthan, India. Curr. Sci. 50 : 33- 34.

Mathur, S., Thakore, B.B.L. and Singh, R.B. (1984) Translocation of systemic fungicides in sunflower and their curative effect on rust caused by *Puccinia helianthi* Schw. Pesticides 18 : 51-52.

Mathur, S., Thakore, B.B.L., Singh, R.B. and Chakravarti, B.P. (1982) Screening sunflower germplasm against rust (*Puccinia helianthi* Schw.) diseases. Sci. Cult. 28 : 213-214.

Mayee, C.D., Sondge, V.D. and Rodge, R.P. (1985) Spray schedule decision model for rust management in groundnut. J. Maharastra Agric. Univ. 10 : 159-161.

McDonald D., Subrahmanyam, P., Gibbons, R.W. and Smith, D.H. (1985) Early and late leaf spots of groundnut. Info. Bull. No. 21 (i-iv) : 1-19, Patencheru 502 324, A.P.

Mehan, V.K., McDonald, D. and Ramakrishna, N. (1986) Varietal resistance in peanut to aflatoxin production. Peanut Sci. 13 : 7- 10.

Mehan, V.K., McDonald, D. and Subrahmanyam P. (1985) Bacterial wilt of groundnut : Control with emphasis on host plant resistance. *In* : Bacterial wilt disease in Asia and South Pacific Canberra, ACT Australian Centre for International Agric. Dis., 112-119 pp.

Melero-vara, J.M., Garcia-baudin, C. and Lopezherrera, C.J. (1982) Control of sunflower downy mildew with metalaxyl. Plant Dis. 66 : 132-135.

Mishra, R.P., Kushwaha, U.S., Khare, M.N. and Chand, J.N. (1972) Rhizopus rot of sunflowers (*Helianthus annus* L.) in India. Indian Phytopath. 25 : 236-239.

Mishra, R.P., Singh, B.P. and Joshi, L.K. (1973) Pod susceptibility of different varieties of til. (*Sesamum orientale*) to *Macrophomina phaseoli* (Maubl). Ashby. JNKVV. Res. J. 7 : 288-289.

Mohanty, N.N. (1958) Cercospora leaf spot of sesame (*Sesamum orientale* L.). Indian Phytopath. 11 : 186-187.

Mohanty, S.S. and Rath, G.C. (!987) Parasitic adoption of seed of *Cuscuta.* Indian Phytopath. 40 : 413-415.

Moniz, L., Syed, G.M. and Raj, H.H. (1968) *Xanthomonas guizotiae* var. *indicus* causing leaf spot disease of *Guizotia abyssinica* Cass. Indian J.Microbiol. 8 : 263.

Moses, G.J. and Reddy Ranga, R. (1989) New viral and fungal diseases on groundnut in the southern Telangana zone of Andhra Pradesh, India. Internatl. Arachis News. (IAN) No.5, May 1989, 10-11 pp.

Moss, M.O. (1972) Aflatoxins and related mycotoxins In 'Phytochemical Ecology' (J.B. Arborne, Ed.) Academic press, London, 121-144 pp.

Muheet, A. and Chauhan., L.S. (1975) Control of phyllody of sesamum (*Sesamum orientale* L.) Madras Agric. J. 62: 219- 220.

Mukewar, P.M., Lambat, A.K., Nath, R., Majumdar, A., Rani, I. and Chandra, K.J. (1974) Blight disease of sunflower caused by *Alternaria helianthi* (Hansf.) Tubaki and Nishihara in India. Curr. Sci. 43: 346-347.

Mulekar, V.G., Mayee, C.D. and Patil, M.A. (1988) Efficacy of a new formulation against *Plasmopara helstedii.* Indian Phytopath. 41 : 496-497.

Muthusamy, M. and Subramanian, N. (1985) A new mycoplasma disease in peanut (*Arachis hypogaea* L.) Curr. Sci. 54 : 1193-1194.

Narain, U. and Chauhan, L.S. (1981) Leaf spot of sunflower caused by *Alternaria tenuissima.* FAO Plant prot. Bull. 29 : 29.

Narain, U. and Koul, A.K. (1984) Myrothecium leaf spot of sunflower. FAO Plant Prot. Bull. 32 : 146.

Natrajan, S., Narayanasamy, P. and Kandaswamy, T.K. (1983) Control of post-emergence root-rot and collar-rot diseases of groundnut. *In* proceedings of the National Seminar on management of diseases of oilseed crops, Madurai, India, 29-30 pp.

Nayak, M.L. and Sharma, R.K. (1981) A new weed host of bacterial blight of sesamum. Indian Phytopath. 33 : 482.

Nolt, B.L., Rajeswari, R., Reddy, D.V.R., Bharathan, N. and Manchar, S.K. (1988) Indian Peanut clump virus isolates : host range, symptomatology, serological relationships, and some physical properties. Phytopathology 78 : 310-313.

Nolt, B.L. and Reddy, D.V.R. (1985) First report of *Polymyxa graminis* in India. Plant Dis. 69 : 542.

Pachkhede, A.U., Vyawajare, S.V. and Shreemali, J.L. (1985) Two new species of *Phoma* from India. Curr. Sci. 54 : 485-486.

Patel, D.J., Patel, B.A. Chavda, J.C., and Patel, H.V. (1988) Record of *Meloidogyne javanica* on groundnut in Gujarat. India. International Arachis Newsletter No. 3 : 16-17.

Patel, H.R., Vaishnav, M.U. and Dhruj, I.U. (1986) Efficacy of aldicarb sulfone and carbofuran flowable seed treatment on plant growth against *Meloidogyne arenaria* on groundnut. Pesticides 20 : 29-31.

Patel, H.R., Vaishnav, M.U. and Dhruj, I.U. (1988) Screening of groundnut lines for resistance to root knot nematode. International Arachis News letter No.3 : 11.

Patel, M.K., Abhyankar, S.G. and Kulkarni, Y.S. (1949) Black rot of Cabbage. Indian Phytopath. 2 : 58-61.

Patel, J.J., Patel, M.S. and Kanzaria, M.V. (1982) Effect of lime and sulphur on iron availability under different soil moisture regimes. Indian J. Agric. Res. 16 : 83-86.

Patil, M.B. and Rane, M.S. (1982) Incidence and control of sclerotium wilt of groundnut. Pesticides 16 : 23-24.

Prasad, B., Verma, O.P. and Daftari, L.N. (1985) Effect of leaf curl disease on seed and oil quality of sesamum (*Sesamum indicum* L.). Curr. Sci. 54 : 388-389.

Purohit, S.D., Ramawat, K.G. and Arya, H.C. (1978) Light microscopic detection of Mycoplasma-like organisam (MLO) in sesamum phyllody. Curr. Sci. 47 : 866-867.

Rani, V.U., Lengam, T.S. and Thirupathaiah, V. (1984) Bending of stem in sesame - a new symptom caused by *Alternaria alternata*. Indian J. Mycol. Plant Pathol. 14 : 98-99.

Rao, G.N. and Rajgopalan, K. (1977) A new record of *Alternaria* causing leaf spot of sunflower. Curr. Sci. 46: 750-751.

Rao, N.N.R. and Pavgi, M.S. (1975) New sclerotial disease of crops in Varanasi. Indian Phytopath. 28 : 401-404.

Rao, R.D.V.J.P., Chakravarty, S.K. and Reddy, A.S. (1988) First report of peanut stripe virus from India. Plant Dis. 72 : 912.

Rao, R.D.V.J.P., Ragunathan, V., Rao, M.V. and Joshi, N.C. (1979) The effect of tomato spotted wilt virus on yield of groundnut. Indian J. Plant Prot. 7 : 161-164.

Rao, Y.P. (1962) Bacterial blight of sesamum (*Sesamum orientale* L.). Indian Phytopath. 15 : 297-298.

Rao, Y.P. and Durgapal, J.C. (1966) Seed transmission of bacterial blight disease of sesamum (*Sesamum orientale* L.) and eradication of seed infection. Indian Phytopath. 19 : 402- 403.

Rathaiah, Y. (1983) Phytophthora blight of sesamum - new to Assam. J. Res. Assam. Agric. Univ. 4 : 69-73.

Rathaiah, Y. and Pavgi, M.S. (1973) Perpetuation of species of *Cercospora* and *Ramularia* parasitic on oilseed crops. Ann. Phytopath. Soc., Japan 39 : 103-108.

Rathnaswamy, R., Sundaram, N., Vindhiyavarman, P., Muthusamy, M. and Ramalingam. R.S. (1986) Groundnut lines resistant to late leaf spot and rust, developed through hybridization of 'Mutant-1' *Arachis hypogaea* Linn. with *A. villosa* Benth. Indian J. agric. Sci. 56 : 537-539.

Raut, J.G. (1985) Effect of charcoal rot caused by *Macrophomina phaseolina* on sunflower plants. Indian Phytopath. 38 : 345-346.

Raut, J.G. (1985 a) Location of *Alternaria helianthi* in sunflower seed and its transmission from seed to plant. Indian Phytopath. 38 : 522.

Raut, J.G. and Bhombe, B.B. (1986) Efficacy of some fungicides and hot water in the control of seed borne infection of *Macrophomina phaseolina* in sunflower. Indian phytopath. 36 : 294-296.

Raut, J.G. and Kale, K.B. (1976) Reaction of sunflower strains to powdery mildew. J. Maharastra Agric. Univ. 1 : 317.

Reddy, D.V.R., Nene, Y.L. and McDonald, D. (1986) Virus disease problems of groundnut, chickpea and pigeonpea in Asia. Trop. Agric. Res. Series No. 19: 67-75.

Reddy, M.N. and Rao, A.S. (1980) Damping-off of groundnut caused by the species of *Fusarium*. Indian J. Mycol. Plant Pathol. 10 : 65-66.

Reddy, M.N. and Rao, A.S. (1983) Some factors in the development of resistance of groundnut seedling to damping-off caused by *Rhizoctonia solani*. Acta Phytopathol. Acad. Sci. Hung. 18 : 37-43.

Reddy, P.C. and Siradhana, B.S. (1978) A new leaf spot disease of sunflower in India. Plant Dis. Reptr. 62 : 508.

Roy, A.K. (1965) Outbreaks and new records, occurrence of sesame powdery mildew caused by *Oidium erysiphoides*. FAO Plant Prot. Bull. 13 : 42.

Saharan, G.S. and Singh, B.M. (1976) Brown leaf spot and blight of sunflower caused by *Septoria helianthi*. Indian J. Mycol. Plant Pathol. 6 : 181.

Saksena, H.K., Kaul, A.K., and Narain, U. (1976) Relative efficiency of fungicides *in vitro* against *Botryodiplodia theobromae* Pat. causing disease of sunflower. Indian J. Farm Sci. 4 : 113.

Samuel, G.S., Govindaswamy, C.V. and Vidhyasekaran, P. (1971) Studies on the Alternaria blight disease of gingelly. Madras Agric. J. 58 : 882-886.

Satya, H.N. (1964) A new species of *Pestalotiopsis* on *Arachis hypogaea* L. from Bhopal. Curr. Sci. 33: 57.

Sehgal, S.P. and Prasad, N. (1966) studies on the penetration and survival of sesamum - phytophthora. Indian Phytopath. 19 : 173-177.

Sehgal, S.P. and Prasad, N. (1972) Note on testing of different varieties, strains and species of sesamum for resistance to Phytophthora blight. Indian J. agric. Sci. 42 : 122.

Seshadri, K. (1976) Effect of fungicides in the control of groundnut rust. Madras Agric. J. 63 : 416.

Shanmugam, N., Natrajan, S. and Ramakrishnan, G. (1976) Fungicidal control of powdery mildew of *Sesamum indicum* L. Madras Agric. J. 58 : 30-32.

Sharifi - Tehrani, A. (1974) Effect of seed disinfection to control sunflower foot rot caused by *Sclerotinia sclerotiorum* (Lib.) de By. Iranian J. Plant Path. 10 : 45-48, 23-24.

Sharma, N.D. (1974) A previously unrecorded leaf spot of peanut caused by *Phomopsis* sp. Plant Dis. Reptr. 58 :640.

Sharma, R.C. and Bhowmik, T.P. (1982) Comparative efficacy of various seed dressing fungicides against Macrophomina root rot of groundnut. Pesticides 16 : 31-32.

Sharma, S.M. (1982) Improved Technology for sesamum and niger. Indian Fmg. 32 : 72-77.

Shekhawat, P.S., Patel, V.N. and Patel, J.G. (1987). Economic fungicidal spray schedule for control of tikka and rust diseases of rainfed groundnut. Indian J. Mycol. Pt. Pathol. 17 : 11-16.

Siddaramaiah, A.L., Kulkarni, S. and Basavarasaiah, A.B. (1979) Occurrence of a new collar rot disease of niger (*Guizotia abyssinica* Cass.). in India. Curr. Sci. 48 : 74.

Singh, A.K., Chakrabarti, D.K. and Basuchaudhary, K.C. (1975) Two new diseases of safflower from India. Curr. Sci. 44 : 397- 399.

Singh, G.R. and Kang, M.S. (1981) Influence of systemic and non- systemic fungicides on the growth and tikka disease of groundnut. III International symposium on Plant Pathology, New Delhi, 170 PP. (Abstr.).

Singh, R.N. (1970) Integrated control of bacterial diseases of sesamum in India. Indian Phytopath. 23 : 155 (Abstr.).

Sivaprakasam, K., Piliayarswamy, K. and Rajaram, S. (1975). Effect of NPK on root rot disease incidence in sunflower. Madras Agric. J. 62 : 308.

Sokhi, S.S. and Jhooty, J.S. (1986) Possible annual recurrence of groundnut rust in Punjab. Indian Phytopath. 39 : 441-443.

Sreenivasulu, P. and Nayudu, M.V. (1986) Assessment of economic yield loss in peanut green mosaic virus (PGMV) infected groundnut. Indian Phytopath. 36 : 403-405.

Sreenivasulu, P., Iizuka, N., Rajeshwari, R., Reddy, D.V.R. and Nayudu, M.V. (1981) Peanut green mosaic virus- a member of the potato virus Y group infecting groundnut (*Arachis hypogaea*) in India. Anal. appl. Biol. 98 : 255-260.

Subrahmanyam, P. and McDonald, D. (1980) Occurrence of *Olpidium brassicae* on groundnut. Trans. Br. Mycol. Soc. 15 : 506-509.

Subrahmanyam, P., McDonald, D. and Gibbons, R.W. (1982) Variation in *Cercosporidium personatum* symptoms on certain cultivars of *Arachis hypogaea*. Oleagineux 37 : 63-68.

Subrahmanyam, P., Moss, J.P., McDonald, D., Rao, P.V.S. and Rao, V.R. (1985) Resistance to leaf spot caused by *Cercosporidium personatum* in wild *Arachis* species. Plant Dis. 69 : 951-954.

Tandon, I.N. and Singh, B.K. (1968) Chemical control of leaf spot of groundnut. Indian Phytopath. 21 : 281-288.

Thakore, B.B.L., Mathur SNEH, and Singh, R.B. (1987) Management of leaf spots and rust disease of peanut by single fungicide. Korean J. Plant. Prot. 26 : 113-115.

Tiwari, R.K.S., Vyas, S.C. and Shastry, P.P. (1987) The role of adjuvants in the foliar absorption of benzimidazoles and in controlling leaf spots of groundnut. Indian J. Plant Prot. 15 : 71-75.

Ukey, R.C. and Mehta, B.K. (1984) Reaction of bunch, spreading and semi-spreading types of groundnut to Phyllosticta leaf spot. J. Maharastra Agric. Univ. 9 : 222-223.

Vajavat, R.M. and Chakravarti, B.P. (1978) Survival of *Pseudomonas sesami* and effect as an antagonistic bacterium isolated from seeds on the control of the disease in the field. Indian Phytopath. 31 : 286-288.

Vander Zijden, A.S.M., Koelensmid, W.A.A.B., Boldingh, J., Barsett, C.B., Ozd, W.O. and Philip, J. (1962) Isolation in crystalline form of a toxin responsible for turkey diseases. Nature 195 : 1060-1062.

Vasudeva, R.S. and Sahambi, H.S. (1955) Phyllody of sesamum (*Sesamum orientale* L.). Indian Phytopath. 8 : 124-129.

Verma, O.P. and Daftari, L.N. (1974) Amount of seed borne inoculum of *Macrophomina phaseoli* and its effect on mortality and growth of sesamum seedlings. Indian Phytopath. 27 : 30-31.

Virk, K.S. and Grover, R.K. (1975) Fungitoxic studies with thiophanate methyl *in vitro* and *in vivo*. Indian J. Mycol. Plant Pathol. 5 : 79-85.

Vyas, S.C. (1981) Diseases of sesamum and niger in India and their control. Pesticides 15 : 10-15.

Vyas, S.C., Prasad, K.V.V., Shastry, P.P. and Shukla, B.N. (1986) Field control of leaf spots and rusts of groundnut by fungicides. Indian Phytopath. 39 : 447-448.

Vyas, S.C., Shastry, P.P., Shukla, B.N. and Varma, R.K. (1985) Two new leaf blight diseases of groundnut. FAO Plant Prot. Bull. 33 : 121-122.

Wangikar, P.D., Buldeo, A.N., Wankar, B.N., Wadibhasme, S.S. and Khune, N.N. (1985) Chemical control of rust of groundnut. PKV Res. J. 9 : 89-94.

Yadav, B.S. and Pathak, A.K. (1985) Nematode pests of oilseed crops and screening of certain varieties against nematodes. *In* Oilseed production, constraints and opportunities. Oxford and IBH publishing company, New Delhi, 479-486 pp.

Dryland Resources & Technology, 7 (1992)
Editors : L.L. SOMANI, R.L. SHYAMPURA
& M.C. MUNDRA


# Towards Plant Type Concept in Moth Bean, (*Vigna aconitifolia*)

*M. M. Bhandari*[*]
[*]Bio Energy Centre, C.T.A.E. Udaipur

## ABSTRACT

Significant phenotypic correlations and sizeable $r_g$ values between grain yield per plant and other characters suggest that selection for high straw yield per plant, more number of effective branches, late maturity, more number of pods per plant, late flowering, long pods and high test weight, would result in plant types possessing high grain yield. Genotypic association of grain yield per day per plant for grain yield per plant, straw yield per plant, pod length and days to maturity is moderate to high. Harvest index is associated with erect plant habit and less straw yield per plant.

For grain yield per plant as dependent variable, 77.1% of the total variation is explained by straw yield, plant habit, pods per plant hundred seed weight, pod length, days to maturity and seeds per pod. For harvest index as dependent variable, 18.19% of the total variation is explained by plant habit, number of effective branches and pods per plant. The later two characters have negative contribution to harvest index. It is suggested that these characters may be given due importance in selection programmes.

Genetic advance and expected gain from different combination of variables for path analysis (significant correlations) have been presented. The results reveal maximum genetic advance (10.09) and expected gain (497.70%) when all the ten significant correlated variables are included in selection programme for grain yield per plant. Inclusion of only three characters-plant habit, days to maturity and straw yield per plant maintained appreciable gains of selection in terms of genetic advance (9.90) and expected gain (486.44%). Results reveal that inclusion of only two characters- plant habit and straw yield per pland does not much reduce the selection gains in terms of genetic advance (9.80) and expected gain (480.79%).

## INTRODUCTION

Moth bean, *Vigna aconitifolia* (Jacq.) Marechal (synonym : *Phaseolus aconitifolius* Jacq.) is an American originated species endemic to India. It has its origin in Tehuacan Valley (Mexico) about five thousand years ago and is grown mostly in India and to a small extent in China.

Information on correlatio, regression and path coefficients of characters in moth bean, which can possibly be used in genetic amelioration of grain yield by construction of a suitable crop ideotype for maximization of crop yield is meagre. The work carried at Rajasthan Agricultural University (Bhandari, 1990a, 1990b, 1991 a, 1991 b, 1991 c, 1991 d, Bhandari & Mehrotra, 1990) has been used to define suitable plant types of moth bean for high grain yield per plant, grain yield per day per plant and harvest index.

## CORRELATION BETWEEN GRAIN YIELD PER PLANT AND OTHER CHARACTERS

Significant and positive phenotypic correlations of grain yield per plant have been obtained for all the characters (Table 1).

The genotypic correlations of grain yield per plant (Table 1) are high with straw yield per plant (0.91), and number of effective branches (0.77); and moderate with days to maturity (0.72), pods per plant (0.66), days to fifty per cent flowering (0.65), pod lenght (0.60), and hundred seed weight (0.54).

Singnificant phenotypic correlations and appreciable $r_g$ values between grain yield per plant and other characters suggest that selection for high straw yield per plant, more number of effective branches, late maturity, more number of pods per plant, late flowering, long pods and more hundred seed weight, would result in plant types possessing high grain yield per plant. Association of the above mentioned characters with grain yield per plant

Table 1 : Phenotypic and genotypic correlation coefficients

| Character | GYPP | | HI | | GYP DPP | |
|---|---|---|---|---|---|---|
| | Pheno | Geno | Pheno | Geno | Pheno | Geno |
| PHb | 0.14** | 0.19 | 0.35** | 0.65 | 0.14** | 0.20 |
| PHt | 0.27** | 0.49 | –0.08 | –0.13 | 0.25** | 0.46 |
| EB | 0.63** | 0.77 | 0.10* | –0.16 | 0.54** | 0.70 |
| DF | 0.56** | 0.65 | –0.10* | –0.14 | 0.34** | 0.43 |
| DM | 0.63** | 0.72 | –0.10 | –0.13 | 0.39** | 0.50 |
| PP | 0.58** | 0.66 | –0.13** | –0.23 | 0.58** | 0.70 |
| PL | 0.48** | 0.60 | –0.05 | –0.03 | 0.46** | 0.61 |
| SP | 0.35** | 0.48 | 0.04 | 0.08 | 0.32** | 0.47 |
| HSW | 0.46** | 0.54 | –0.07 | –0.09 | 0.33** | 0.41 |
| SYPP | 0.82** | 0.91 | –0.29** | –0.40 | 0.74** | 0.95 |
| GYPP | 1.00 | 1.00 | 0.22** | –0.05 | 0.95** | 0.85 |
| HI | 0.22** | –0.05 | 1.00 | 1.00 | 0.30** | –0.05 |
| GYP DPP | 0.95** | 0.95 | 0.30** | –0.05 | 1.00 | 1.00 |

* and ** significant at 0.05 and 0.01 level respectively.

The characters abbreviated are plant habit, plant height, number of effective branches, days to 50% flowering, days to 75% maturity, number of pods per plant, pod length, number of seeds per pod, hundred seed weight, straw yield per plant, grain yield per plant, harvest index, grain yield per day per plant.

elucidates the importance of proper source and sink relationship. The sink potential assumes importance in realizing the greater yield potential of the end product–the grain. The postulated relationship between source and sink-that is, the effect of sink size (number and potential size of reproductive organs; in the present case number of pods, length of pods, and hundred seed weight) upon the production of photosynthates (straw yield per plant, number of effective branches and days taken to flowering in the present case) and the rate and time available for transport (days taken to maturity from flowering in the present case) to storage organs-is important for realization of greater grain yield per plant (Eastin *et al.*, 1969). High yields from a variety can be obtained only when enough photosynthetic products are first produced and then accumulated in greater extent in the form of grain. The photosysthetic efficiency of variety thus becomes source as producer of photosynthetic product; efficient distribution of this product- is the sink. For exploitation of the yielding potentiality of a variety the relative size of source and sink is required to be increased as high yielding potential can only be realised when a balanced increase has been achieved both in the source and the sink.

## CORRELATION BETWEEN GRAIN YIELD PER DAY PER PLANT AND OTHER CHARACTERS

The phenotypic correlations of grain yield per day per plant are positive and highly significant with all other characters. Positive genotypic correlations are obtained for grain yield per day per plant for all the characters except harvest index. Genotypic correlations are high for grain yield per plant (0.95) and straw yield per plant (0.87); and moderate for pods per plant (0.70), effective branches (0.70), pod length (0.61) and days to maturity (0.50).

Highly significant phenotypic correlation coefficients and appreciable $r_g$ values between grain yield per day per plant and other characters reveal that selection for grain yield per day per plant should be made through its indicative characters viz., high grain yield per plant, high straw yield per plant, more number of effective branches, more number of pods per plant, long pods and late maturity.

Although the same nine characters serve as indicators for grain yield per plant as well as for grain yield per day per plant, yet $r_g$ values of pods per plant and hundred seed weight appreciably increase (values for pod length and plant habit also somewhat increase) when these characters are used as indicators of grain yield per day per plant. The $r_g$ values of grain yield per day per plant considerably decrease for days to fifty per cent flowering, days to maturity and number of effective branches. The respective increase and decrease in correlation of these indicative characters are self explanatory in that these would bring high yielding short duration varieties, and as such instead of absolute yield, the per day productivity would be higher.

## CORRELATION BETWEEN HARVEST INDEX AND OTHER CHARACTERS

Harvest index exhibits significant positive phenotypic correlations (Table 1) with plant habit (0.35) and grain yield per plant (0.22). It shows negative phenotypic correlations with straw yield per plant (-0.29). pods per plant (-0.13), days to flowering (-0.1) and days to maturity (-0.1).

The genotypic association of harvest index is moderate and positive with plant habit (0.65); and low and negative with straw yield per plant (-0.40), and pods per plant (-0.23). Hence erect plant type with less foliage should be selected for increased per hectare yield (Fig 1 & 2).

For increasing per hectare grain yield, more number of productive but non-competitive plants will have to be fitted per unit area (Bhandari, 1976). Increase in plant density can be feasible only with plant types possessing high harvest index. Present study reveals that high harvest index in moth bean is associated at genotypic level with erect plant habit, less straw yield per plant and fewer number of pods per plant. The association of spreading plant types with reduced harvest index make them suitable for high fodder yields only.

Fig. 1. Erect plant type of moth bean.

Fig. 2. Spreading plant type of moth bean

## GRAIN YIELD PER PLANT - REGRESSION COEFFICIENTS

Regression analyses of ten idependent variables (before backward elimination) and seven (after elimination) important characters are presented in Table 2. Unstandardized regression analysis revealed that partial regression coefficients are highly significant for straw yield per plant, pods per plant, plant habit, hundred seed weight, pod length and seeds per pod, and non-significant for other five characters. These ten characters explain 77.25 per cent of the variation in grain yield per plant as shown by 100 R-square (coefficient of determination) value. After dropping non-significant variables one by one with backward elimination procedure, reconstructed multiple regression equation explains 77.09 per cent of the total variation in grain yield per plant. Thus dropping three variables reduces the explained variation by 0.16 per cent only.

As standardized Beta values are directly comparable, multiple regression (standardized) analysis is done after dropping non-significant variables one by one with backward elimination procedure (Table 2). Following multiple regression equation has been constructed :

$$\begin{aligned} Y - Y = {} & 0.1520\,(X_1 - \overline{X}_1) + 0.0915\,(X_5 - \overline{X}_5) \\ & + 0.1478\,(X_6 - \overline{X}_6) + 0.1174\,(X_7 - \overline{X}_7) \\ & + 0.0741\,(X_8 - \overline{X}_8) + 0.1253\,(X_9 - \overline{X}_9) \\ & + 0.5521\,(X_{10} - \overline{X}_{10}) \end{aligned}$$

According to Beta values (Table 1), straw yield is the most important contributor followed by plant habit, pods per plant, hundred seed weight, pod length, days to maturity and seeds per pod. Therefore, it is suggested that these seven characters may be given due importance in selection programmes. Straw yield is the most important character contributing towards grain yield. This elucidates the source-sink relationship. Straw yield accounts for total photoshynthesizing capacity of the plant. Adequately developed source (Photosynthesizing organs) may usually reflect in adequate filling of sink (the grain). So maximum change in grain

Table 2 : Regression analysis of ten independent and seven important characters on grain yield per plant

| Character | Before eliminetion | | After elimination | | Percentage contribution to $R^2$ Value | S. E. (b) | T Value |
|---|---|---|---|---|---|---|---|
| | Standardized regression (B value) | Regression coefficient (b value) | Standardized regression coefficient (B value | Regression coefficient (b value) | | | |
| $X_{10}$ | 0.5353 | 0.1197** | 0.5521 | 0.1235** | 59.59 | 0.01 | 12.78 |
| $X_6$ | 0.1425 | 0.0204** | 0.1478 | 0.0212** | 11.24 | 0.005 | 4.36 |
| $X_1$ | 0.1419 | 1.0253** | 0.1520 | 1.0985** | 2.93 | 0.15 | 7.20 |
| $X_9$ | 0.1194 | 1.0638** | 0.1253 | 1.1160** | 7.65 | 0.24 | 4.68 |
| $X_7$ | 0.1159 | 0.9267** | 0.1174 | 0.9387** | 7.55 | 0.21 | 4.48 |
| $X_5$ | 0.0902 | 0.0336 | 0.0915 | 0.0340** | 7.64 | 0.012 | 2.97 |
| $X_8$ | 0.0741 | 0.3154** | 0.0741 | 0.3153** | 3.40 | 0.11 | 2.88 |
| $X_3$ | 0.0553 | 0.0719 | | | | | |
| $X_2$ | –0.0140 | –0.0106 | | | | | |
| $X_4$ | –0.0125 | –0.0076 | | | | | |

Multiple correlation coefficient, R = 0.8789**; 0.8780**

Coefficient of determination 100 $R^2$ = 77.25; 77.09

Intercept with S.E. = –9.7911 ± 0.3929; –10.4801 ± 0.3849

* Significant at 5% level

** Significant at 1% level.

yield is possible with a unit change in straw yield in comparison to other determinants of grain yield. The plant habit (erect-spreading) accounted for competition or reduction in competition between and within plants (between organs of the same plant). However its direct contribution to R- square value was less & could have contributed indirectly via other characters. Pods per plant, hundred seed weight and seeds per pod are yield components on which grain yield per plant can be mathematically calculated or directly predicted, so there positive and high mangitude of association with grain yield in moth bean is self explainatory. Although many of the prevalent long duration moth varieties yield high grain yield per plant in accordance with results of regression analysis yet its small direct contribution to R-square value reveals scope for developing high yielding short duration moth bean varieties.

## HARVEST INDEX - REGRESSION COEFFICIENTS

Regression analysis of nine independent characters (before elimination) and three important characters (After elimination) are presented in Table 3. Unstandardized regression analysis (before elimination) revealed that partial regression coefficients were highly significant for plant habit and pods per plant, significant for number of effective branches and nonsignificant for other six characters. Percentage of variation in harvest index explained by these nine characters as shown by coefficient of determination is 18.35.

After dropping non-significant variables one by one with backward elimination procedure reconstructed multiple regression (unstandardized coefficients) equation (Table 3) explains 18.19 % of the total variation in harvest index. Thus dropping of six variables reduces the variation explained by 0.16 per cent only. According to standardized Beta values (Table 3), plant habit is the most important contributor followed by number of effective branches and pods per plant out of the nine characters studied. After dropping non-significant variables one by one with backward elmination procedure, multiple regression (standardized coefficients) equation is constructed as follows :

$$Y - \overline{Y} = 0.4027\ (X_1 - \overline{X}_1) - 0.1160\ (X_3 - \overline{X}_3) - 0.1104\ (X_6 - \overline{X}_6)$$

Table 3 : Regression analysis of nine independent and three important characters on harvest index

| Character | Before eliminetion | | After Elimination | | Percentage contribution to $R^2$ Value | S. E. (b) | T Value |
|---|---|---|---|---|---|---|---|
| | Standardized regression (B value) | Regression coefficient (b value) | Standardized regression coefficient (B value) | Regression coefficient (b value) | | | |
| $X_1$ | 0.4094 | 3.2671** | 0.4027 | 3.2140** | 84.38 | 0.30 | 10.72 |
| $X_6$ | –0.1180 | –0.0187** | –0.1104 | –0.0175** | 9.16 | 0.01 | 2.66 |
| $X_3$ | –0.1146 | –0.1644* | –0.1160 | –0.1664** | 6.46 | 0.06 | 2.76 |
| $X_5$ | 0.0368 | 0.0151 | | | | | |
| $X_7$ | –0.0335 | –0.2955 | | | | | |
| $X_8$ | –0.0116 | –0.0544 | | | | | |
| $X_2$ | –0.0063 | –0.0052 | | | | | |
| $X_9$ | –0.0057 | –0.0562 | | | | | |
| $X_4$ | –0.0036 | –0.0024 | | | | | |

Multiple correlation coefficient, R = 0.4284**; 0.4264**

Coefficient of determination 100 $R^2$ = 18.36; 18.19

Intercept with S.E. = 14.1502 ± 0.7925; 13.7436 ± 0.3807

* Significant at 5% level

** Significant at 1% level.

According to this equation, plant habit followed by number of effective branches and pods per plant are three important contributors to harvest index. Regression analysis reveal that harvest index is mainly dependent on plant habit, the effect of which is positive and direct. Effective number of branches and number of pods per plant have negative contribution to harvest index. Erect plant habit should therefore be given considerable importance in selection programme for increasing harvest index. Erect plant types possess high grain to straw ratio; and should form a basis in intensive agriculture for high yielding programme which can further pave way for increase in number of high yielding plants per unit area. Our present findings also reveal that erect plant types with less number of effective branches and low number of pods per plant, should be selected for increasing hrvest index.

## GENETIC ADVANCE AND EXPECTED GAIN FROM PATH ANALYSIS

The results reveal maximum genetic advance and expected gain when all the significant correlated variables are included in selection programme for grain yield per plant. Dropping of one of the two variables-days to flowering or days to maturity does not much affect the results. Therefore any one of these two characters can be included in selection programme. Inclusion of only three characters- plant habit, days to maturity and straw yield per plant maintain appreciable gains of selection in terms of genetic advance (9.90) and expected gain (486.44%). Results further reveal that inclusion of only two characters plant habit & straw yield per plant (and dropping out all other characters) does not much reduce the selection gains in terms of genetic advance and expected gain (Table 4).

Selection for grain yield per day per plant could be very effective when plant habit, plant height, number of branches, days to flowering, pod number, pod length, seeds per pod, hundred seed weight & straw yield per plant are included in the programme (Table 4).

Table 4 : Genetic advance and expected gain from different combinations of variables for path analysis

| Variable for path analysis (First dependent) | Pooled over both environments | |
|---|---|---|
| | Genetic advance | Expected gain (%) |
| 11 01 02 03 04 05 06 07 08 09 10 | 10.09 | 497.70 |
| 11 01 02 03 05 06 07 08 09 10 | 10.09 | 497.69 |
| 11 01 02 03 04 06 07 08 09 10 | 10.08 | 497.38 |
| 11 01 04 05 10 | 9.90 | 486.54 |
| 11 01 05 10 | 9.90 | 486.44 |
| 11 01 04 10 | 9.87 | 484.56 |
| 11 01 10 | 9.80 | 480.79 |
| 11 03 05 10 | 9.71 | 475.10 |
| 11 06 08 09 | 9.22 | 446.03 |
| 11 06 07 | 8.47 | 402.13 |
| 11 06 09 | 8.26 | 389.31 |
| 11 06 08 | 8.22 | 387.28 |
| 12 01 02 03 04 05 06 07 08 09 | 5.39 | 543.33 |
| 13 01 02 03 04 06 07 08 09 10 | 95.77 | 6184.14 |

01 (PHb), 02 (PHt), 03 (EB), 04 (DF), 05 (DM), 06 (PP), 07 (PL), 08 (SP), 09 (HSW), 10 (SYPP), 11 (GYPP), 12 (HI), 13 (GYPDPP).

The contribution of important characters towards grain yield per plant & harvest index has been summarised in Tables 5 & 6.

Table 5 : Grain yield per plant

| Determinant | Genotypic correlation coefficient | Standardized regression coefficient | Direct effect (path coefficient) |
|---|---|---|---|
| Straw yield | 0.91 | 0.5521 | 0.4682 |
| Plant habit | 0.19 | 0.1520 | 0.1533 |
| Pods/plant | 0.66 | 0.1478 | 0.2711 |
| Hundred seed weight | 0.54 | 0.1253 | 0.1919 |
| Seeds/pod | 0.48 | 0.0741 | 0.1837 |

Table 6 : Harvest index

| Determinant | Genotypic correlation coefficient | Standardised regression coefficient | Direct effect (path coefficient) |
|---|---|---|---|
| Plant habit | 0.6488 | 0.4027 | 0.7363 |
| Pods/plant | –0.2292 | –0.1104 | –0.1889 |
| Effective branches | –0.1629 | –0.1160 | –0.3001 |

## RESEARCH GAPS

Breeding moth plant type for high rainfall high fertility conditions based on erect plant types is altogether different from breeding for low rainfall low fertility conditions based on spreading plant types. Hybridization methods utilizing spreading varieties should form the immediate base for exploring moth germplasm for dry land agriculture.

## REFERENCES

Bhandari, M.M. (1976) The effect of inter and intra plant competition on crop yield. Raj. Agric. 8 : 51-55.

Bhandari, M.M. (1990 a) Correlation analysis in moth bean. Trans. ISDT. 15 : 33-42.

Bhandari, M.M. (1990 b) Path analysis in moth bean. Annals of Arid Zone, 29 (2) : 147-148.

Bhandari, M.M. and Mehrotra H.N. (1990) Genetic divergence in moth bean. Annals of Arid Zone, 29 (3) : 227-230.

Bhandari, M.M. (1991 a) Variability in moth bean under two environments. Trans. ISTD. 16 : 57-64.

Bhandari, M.M. (1991 b) Euclidean cluster analysis in moth bean under high fertility conditions. Trans. ISTD. 16 : 101-104.

Bhandari, M.M. (1991 c) Towards plant type concept in moth bean. Seventyeighth session of Indian Science Congress, Indore, 78 : 34-35.

Bhandari, M.M. (1991 d) Comparison of euclidean, D-square & canonical analysis in moth bean. Seventyeighth session of Indian Science Congress, Indore, 78 : 35-36.

Eastin, J.D., Haskins, F.A., Sullivan, C.V. and Van Bavel. eds. (1969) Physiological Aspects of Crop Yield, CSSA and ASA, Madison, Wisconsin, USA.

Dryland Resources & Technology, 7 (1992)
Editors : L.L. SOMANI, R.L. SHYAMPURA
& M.C. MUNDRA


# Agroforestry Research Review with Special Reference to the Semi- Arid Region of India

*A.N. Giri*[*] *and J.C. Hetherington*[**]
[*]AICRP on Weed Control, MAU,
Parbhani–431 402
[**]Univ. College of North Wales, Bangor,
Gwynedd, LL57 2UW (U.K.)

## INTRODUCTION

The agro-forestry practices have been followed in different parts of India for a long time. However, the systematic research work on this aspect has been started recently at some of the locations in almost all the agro-ecological zones. The semi-arid tropics is one of the major zone which constitutes about 29.57 per cent of the country's area. The agro-forestry research work in India with special reference to semi-arid region has been reviewed with an object to find out suitable tree species for initiating silvo- pastoral/agri-silvicultural/silvi-agricultural research work which will enable to develop farming systems that will help to increase and stabilise agricultural/forage/fuelwood/ timber production through better use of natural and human resources. It is also felt essential to review the present and future requirements of the country for foodgrains, forage, fuelwood and timber, which will be useful for planning the agroforestry research work.

India has a land frontier of 15,200 km and its seacoast runs to a length of 6,100 km. The mainland extends from 8° N to 37° N latitude and from 68° E to 97° E longitudes. It measures 3214 km between extreme latitudes and 2933 km between extreme longitudes. It is the second most populous and seventh largest country in the world. Its area is more than 32.8 million ha and population is more than 751 million.

## LAND UTILISATION IN INDIA

The land use pattern as compiled from the statistical books issued by the different states reveal the following (table 1, figure 1) land use pattern in the country (Anon., 1987).

Table 1 : Land Utilisation in India.

| Sr. No. | Land use | Area (million ha) | Percentage of total area |
|---|---|---|---|
| 1 | Agriculture (cultivable land). | 154.70 | 47.0 |
| 2 | Forests (area officially recorded as forests with or without tree cover). | 75.18 | 22.8 |
| 3 | Permanent pastures and other grazing lands. | 12.15 | 3.7 |
| 4 | Land under cultivable tree crop and groves. | 3.91 | 1.3 |
| 5 | Cultivable waste land. | 16.64 | 5.1 |
| 6 | Land under other non-agricultural uses. | 17.53 | 5.3 |
| 7 | Barren and waste land. | 24.60 | 7.5 |
| 8 | Area for which no returns exist. | 24.09 | 7.3 |

Although land uses are classified in to eight categories the major land uses are classified as cultivated land, forest land, pasture land and waste land. 'Waste land' primarily refers to land unfit for agricultural crops, even though the land may be suitable for growing grasses and trees, and it could be termed 'wasted land' instead.

### Present State of Forests in India

In 1987, the recorded forest area in the country is 75.18 million ha which works out at 22.18 per cent of the geographical area of the country. However, as per estimates made by Forest Survey of India and National Remote Sensing Agency, the actual forest cover is 64.20 million ha or 19.52 per cent of the geographical area of the country. Of the total forest cover of 64.20 million ha only 36.14 million ha is of adequate density (over 40 per cent). The effective forest cover is, therefore, limited to 10.88 per cent of the country's geographical area (Anon., 1989).

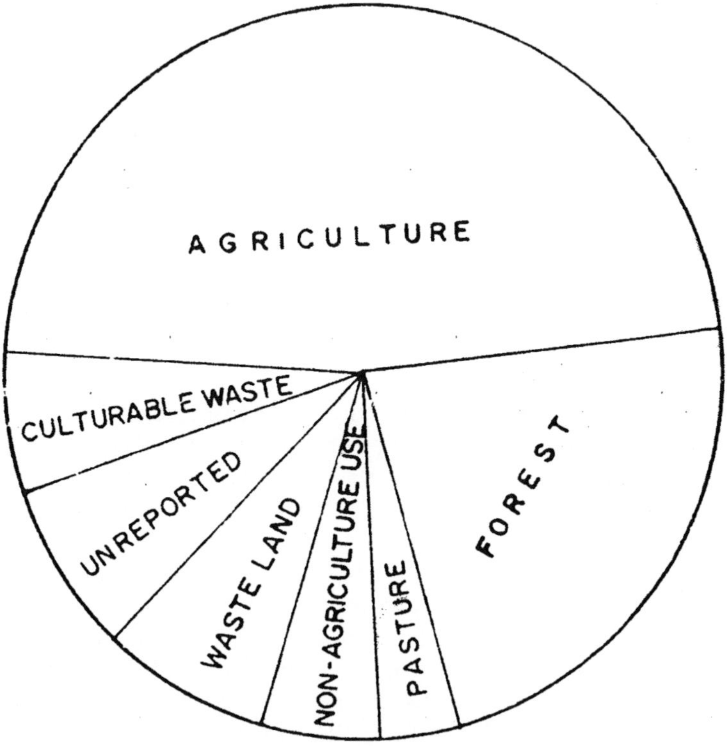


Fig. 1. Land utilisation in India.

## Per Capita Forest Area

The per capita forest area in the country is slightly more than 0.1 ha while per capita agricultural land is more than 0.2 ha. The per capita forest area compares poorly with the world average of 1.0 ha. The world average for per capita agricultural area is 0.3 ha.

The table 2 indicates the Indian situation in comparison to some other countries (Anon., 1987).

Table 2 : Per capita area (ha) in different countries.

| Country | Per capita area in ha | |
|---|---|---|
| | Agricultural land | Forest land |
| India | 0.2 | 0.1 |
| U.S.S.R. | 0.9 | 3.6 |
| United States | 0.9 | 1.3 |
| Japan | 0.04 | 0.2 |
| United Kingdom | 0.1 | 0.04 |
| Thailand | 0.4 | 0.5 |
| Canada | 1.9 | 14.2 |
| Australia | 3.2 | 7.6 |
| Nepal | 0.3 | 0.3 |

## Present State of Fodder

India's cattle population has increased from 292 million in 1950 to 416 in 1985 (Anon., 1987). Presently, only a third of the fodder that is required to feed the country's livestock populaton is available (Singh, 1987). The combined availability of green fodder from permanent pastures, other grazing lands, agricultural lands and forest is estimated at 434 million tonnes (Anon., 1987). The big gap between availability and requirement of green fodder has resulted in unlimited and unrestricted grazing in forest lands. Excessive grazing has degraded forest soils and hampered the natural regeneration. It is estimated that as of now 90 million domestic animals graze in forests. This is far beyond the carrying capacity of the resources which is estimated to be 31 million cattle unit (Anon., 1989). Because of this uncontrolled grazing the productivity of these lands is continuously reduced and the forests are being gradually depleted.

## Present Demand for Fuel/Firewood

Over the years the population in the country has risen rapidly from 361 million in 1950 to 751 million in 1985, while the production from forests has remained static. The two major forest products, besides timber, on which the country depends overwhelmingly are firewood and green fodder. Nearly, 80 per cent of the country's population living in the rural areas depends on fuelwood as the main energy source for domestic use mainly for cooking. The current annual fuel / firewood demand in the country is 235 million cu.m against a production of 40 million cu. m (Anon., 1989).

## Present State of Demand for Timber

The annual demand for industrial wood in the country is 27.58 million cu. m against a production of 12 million cu. m. The average annual production of wood $ha^{-1}$ considering the recorded forest area (75.18 million ha) in the country is 0.7 $m^3$ which compares poorly with the world average of 2.1 $m^3$ (Anon., 1989). The growing stock $ha^{-1}$ of the actual forest cover in the country works out at about 65 $m^3$.

## Presnet State of Erosion Problems

Clearing forest for agriculture is a historical phenomenon. The begining of agriculture was in the form of shifting cultivation. This practice is still followed in North-Eastern region of the country. About 4.35 million ha of forest land is under shifting cultivation and 0.7 million ha of forest land is under encroachment, while 4.328 million ha of forest land were diverted to non-forest use between the period 1951-1980. As a result of this about 44 per cent of the geographical area of India is subjected to serious erosion hazards, causing an estimated loss of 5,334 million tonnes of soil annually, 1,572 million tonnes of which find their way into the sea (Singh, 1987). The National Forest Policy aims that the country should have a minimum of 33 per cent of the total area under forest cover. The mountainous regions should have 2/3 of the area under forest cover to prevent erosion and land degradation (Anon., 1989).

It is estimated that by the year A.D. 2000 the population will be around 1000 million and that of cattle will not be less than 500 million. Similarly, the fuelwood demand by the same year, according to a Planning Commision study, will be 172 million tonnes whereas the Advisory Board on Energy (ABE) estimates the demand will in the range of 300 to 330 million tonnes by 2004/5 (Anon., 1989). Thus, besides the need for increasing food production to feed the increasing population of the country, the urgency of meeting fuel, fodder and timber requirements and preserving the ecological and environmental balance can not be understated. In order to realise the goal of producing 225 million tonnes of food by the year A.D. 2000 in India, an estimated 20 million tonnes of plant nutrients would be required, a substantial part of which will have to be contributed by sources other than chemical fertilisers (Singh, 1987).

Though, India has made massive investments in agriculture and achieved spectacular success in food production, presently its attention is becomming more focused on the problems of the acute shortage of fodder, fuelwood, timber and other forest products. Appropriate agroforestry systems, therefore, are to be developed for each location to achieve the self sufficiency in food, fodder, fuelwood, timber and other forest products, besides preserving the ecological and environmental balance.

## Agroforestry Systems/Practies in India

In India agroforestry is an age-old practice. In every village there are combinations of tree, crop and animal husbandry activities according to local requirements. The woody perennials have long been deliberately grown in conjunction with field crops, spices, horticultural crops and pasture with the objective of meeting various requirements of the farmer's home and farm and also as an insurance against the risk of uncertain weather conditions. Thus these combinations have stood the test of time and have long been practiced. The agroforestry systems namely, shifting cultivation (jhum), taungya system, *Prosopis cineraria* (khejri) - based system, homestead (home gardens), multiple tree planting, growing tea and coffee under shade trees, coconut based farming etc. have all been followed in different parts of the country for a long time. Not only have these systems been widely practiced, but some have also beeen widely researched e.g. growing tea under shade trees since the latter part of the 19th century, fodder/fuel plantation since the 1950's (Tejwani, 1987).

To cope with some of the challenges and tackle the problem of food, fodder and environmental security the potentials of agroforestry need to be fully exploited. Presently, initiatives are being undertaken by the Governments, farmers, nonagricultural organisations and industries to develop appropriate agroforestry systems and popularise them. The present status of forestry/agroforestry research in India is given below.

## FORESTRY/AGROFORESTRY RESEARCH

Forestry research in India started in the later part of the 19th century. The real start of organised forest research was made with the establishment of the Forest Research Institute (F.R.I.) in 1906 at Dehra Dun and it started with six branches namely, Silviculture, Working Plans, Forest Zoology, Forest Botany, Forest Economics and Forest Chemistry. Since then the F.R.I., Dehra Dun has expanded.

The Indian Council of Forest Research and Education created in 1987. In addition to the existing F.R.I., at Dehra Dun new Institutes to deal with regional problems on forestry rese were created during April, 1988. These are :

i) Institute of Rain and Moist Deciduous Forest Re Jorhat

ii) Institute of Forest Genetics and Tree Breeding, C batore.

iii) Institute of Wood Science and Technology, Banglore.

iv) Institute of Arid Zone Forestry Research, Jodhpur.

v) Institute of Deciduous Forests, Jabalpur.

### Organisations Involved in Forestry Research

The following organisations are involved in forestry/agroforestry research (Anon., 1989).

#### Indian Council of Forestry Research and Education

The six Institutes, depending upon regional problems, are carrying out basic as well as applied research on silviculture, Forest Mensuration, Economics and Management, Social Forestry, Forest Ecology, Soils and Environment, Forest Genetics and Tree

Propagation, Forest Protection, Forest Products Utilization, Wood Anatomy, Systematic Botany, Plant Physiology, Minor Forest Products and Forest Operations.

**State Forest Departments**

Management of forests is mainly the responsibility of the states. Field oriented research is carried out by forest departments. Most of the states have one Silviculturist and some states have two or three. Madhya Pradesh, Kerala and Maharashtra states have their own fully-fledged research institutes.

The main studies carried out by state forest departments are :

i) Developing nursery and plantation technology,

ii) Phenological studies,

iii) Fertiliser and spacing trials,

iv) Technology for refractory sites,

v) Monoculture and mixed plantation studies,

vi) Trials of indigenous and exotic species,

vii) Sample plot establishment and growth measurements,

viii) Establishment of preservation plots for ecological studies,

ix) Agroforestry and farm forestry for producing fuel, fodder and small timber.

## Indian Council of Agricultural Research

Recently, research on agroforestry has been taken up for fodder, fuel and minor forest produces. Research is conducted through seven Institutes located in Karnataka, Kerala, Rajasthan, Uttar Pradesh and West Bengal. In addition, the All India Co-ordinated Research Project on Agroforestry was lunched during 1983 at 20 participating centres in the country (see figure 2). Now, this network has been expanded to 31 centres during the seventh Five Year Plan with an object of establishing Agroforestry Research Centres.

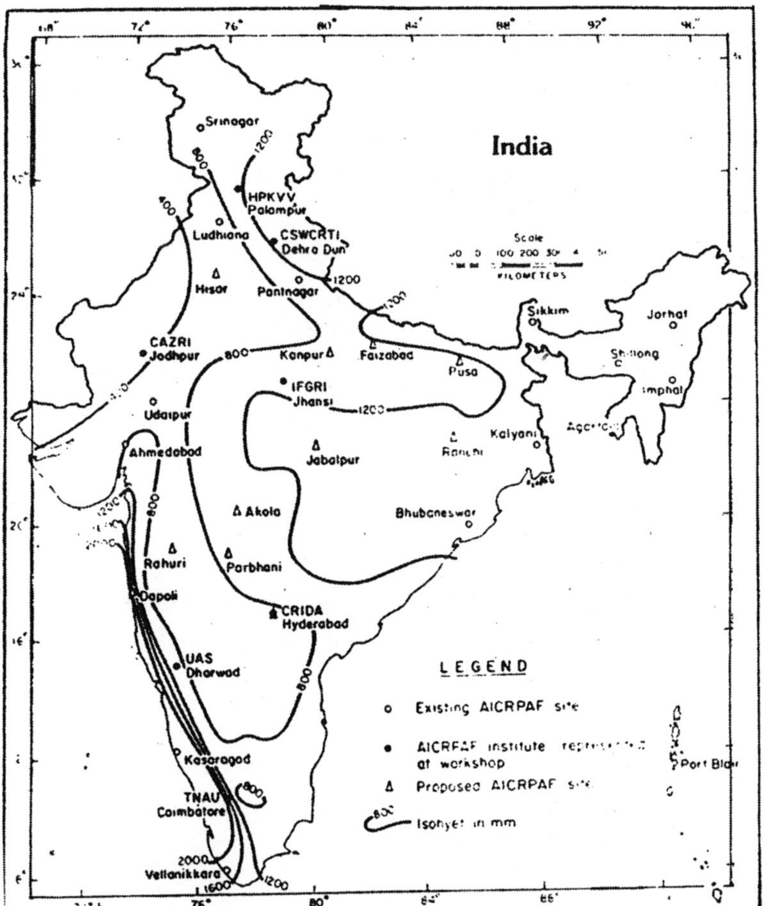


Fig. 2. Location of centres by rainfall zone in the all India coordinated research project on agroforestry.

**Council of Scientific and Industrial Research**

The Regional Research Laboratory, Jorhat (Assam), Central Fuel Research Institute, Dhanbad (Bihar) and National Chemical Laboratory, Pune (Maharashtra) are all working on insecticidal and antifeedant properties from plants, minor forest products, paper and pulp, design and development of biomass gasifier and biotechnological evaluation.

**Agricultural and Other Universities**

Eight agricultural and four other universities have taken up forestry research work. These universities are located in Andhra Pradesh, Assam, Bihar, Gujarat, Haryana, Karnataka, Punjab, Meghalaya, Tamil Nadu and Uttar Pradesh. The main research works conducted by the universities are :

i) Selection of multipurpose tree species,

ii) Field trials on fuelwood species,

iii) Trials on indigenous and exotic species,

iv) Development of silvicultural and management practices,

v) Study of agroforestry systems and investigation of the economical potential of different systems,

vi) Fertiliser trials and biomass production,

vii) Control of weeds through herbicides and

viii) Genetic improvement of trees and the development of tissue culture techniques.

**Other Central/State Organisations**

These include the International Crop Research Institute for Semi- Arid Tropics (ICRISAT, Andhra Pradesh); Indian Institute of Sciences, Banglore ; Indian Plywood Industrial Research Institute, Banglore ; Central Food Technology Research Institute, Mysore ; Council for Science and Technology, Commonwealth Institute of Biological Control, Banglore and Central Drug Research Institute, Lucknow. These Institutes are working on agroforestry studies under rainfed conditions, feeding habits of animals, impact of forest based industries on forests, peeling properties and use of species in plywood making, processing techniques for minor forest produces, control and eradication of weeds through biological methods, tissue culture techniques, alterante sources of energy and the survey, collection and identification of medicinal plants for drugs etc.

### Private Organisations

There are a number of private organisations engaged in forestry research. Bhartiya Agro-Industries Foundation, Pune (Maharashtra) and Tata Energy Research Institute, New Delhi, are the main organisations working on multipurpose fuelwood and nitrogen fixing tree species, studies on petro-crops, energy plantations, mycorrhizae for growth of trees and tissue culture technology.

## Agroecological Zones/ Regions of India

India has been classified into eight agroecological regions. These are as follows :

1. Humid western Himalayan region.
2. Humid Bengal - Assam basin.
3. Humid eastern Himalayan region and Bay islands.
4. Subhumid Sutlaj - Ganga alluvial plains.
5. Subhumid to humid eastern and south - eastern uplands.
6. Arid western plains.
7. Semiarid lava plateau and central highlands.
8. Humid to semiarid Western Ghats and Karnataka plateau.

The agroforestry research work is reviewed with special reference to semi-arid tropics (Zone 7). The major features of the zone are described below.

### Semiarid Lava Plateau and Central Highlands

Singh (1987) has described the region as a plateau ranging from 300 to 1,400 m of horizontally bedded sedimentaries and basalt. Between the highlands, there are wide alluvial plains. The climate is semiarid with extremes of temperature and variable rainfall. The annual precipitation, confined to June - October, varies from 70 to 125 cm except in the Western Ghats where it varies from 335 to 743 cm. The major soil groups of the region are alluvial black and lateritic, mixed red and black and yellowish brown (Paleustalfs, Rhodustalfs, Haplustalps, Chromusterts, Plinthudults and Haplustults). The dominant forest types include moist and dry deciduous, semievergreen thorny and scrubby vegetation. Semi-evergreen forests are confined to the western Ghats and the rest of the plateau is covered with deciduous types. Low and uncertain rainfall, undulating topography and inadequate irrigation impede field operations.

The distribution of semi-arid areas has been tabulated in table 3 and depicted in figure 3.

Table 3 : Distribution of semi–arid areas.

| State | Semi–arid area ($km^2$) | Percentage |
|---|---|---|
| **Hot** | | |
| Rajasthan | 1, 21, 020 | 13 |
| Gujarat | 90, 520 | 9 |
| Punjab and Haryana | 58, 650 | 6 |
| Maharashtra | 1, 89, 580 | 19 |
| Karnataka | 1, 39, 360 | 15 |
| Andhra Pradesh | 1, 38, 670 | 15 |
| Tamil Nadu | 95, 250 | 10 |
| Uttar Pradesh | 64, 230 | 7 |
| Madhya Pradesh | 59, 470 | 6 |
| **Cold** | | |
| Jammu and Kashmir | 13, 780 | |
| **Total** | 9, 70, 530 | |

The total area of the zone constitute about 29.57 per cent of the Country's area (Mann and Muthana, 1984).

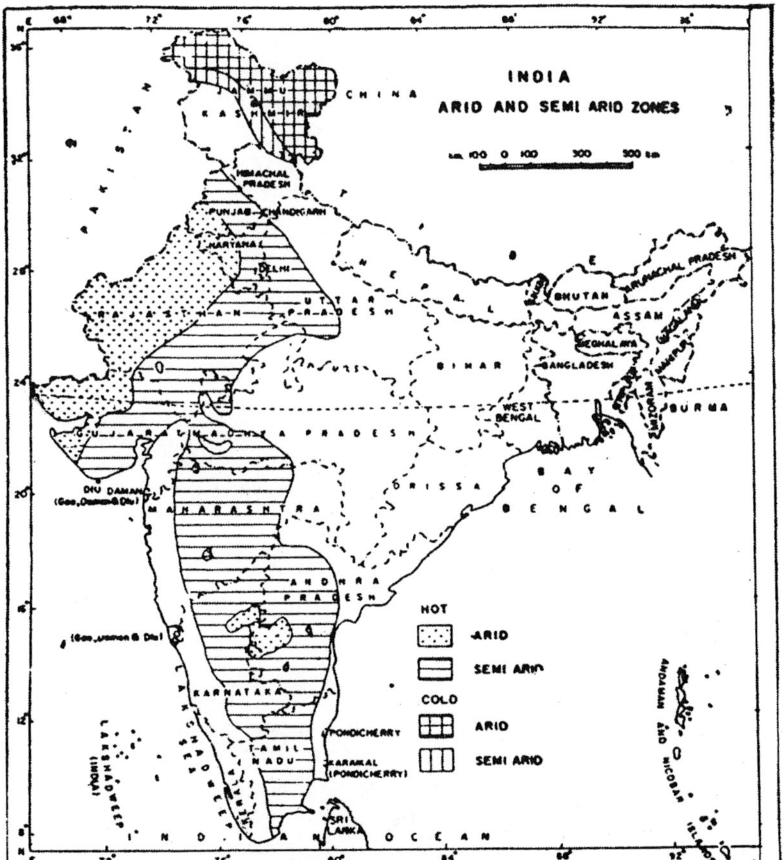


Fig. 3. Distribution of semi-arid areas.

## AGROFORESTRY RESEARCH REVIEW

### Studies on Tree / Shrub Species

Raddi (1981) stated that *Cenchrus ciliaris, Stylosanthes hamata, S. scabra* and the fodder trees *Leucaena leucocephala, Hardwickia binata, Albizzia lebbek, Azadirachta indica* and *Dalbergia sissoo* are suitable for cultivation in the degraded.forest areas in low rainfall regions of Maharashtra. Trials of tree species (exotic and indigenous) conducted at four range management areas of western Rajasthan revealed that *Acacia tortilis, Dichrostachys glomerata, Colophospermum mopane, Azadirachta indica* and *Leucaena leucocephala* performed well in all the areas of the region and are considered as desirable species for the semi-arid zone of Rajasthan (Mann and Muthana, 1984).

Bose and Bandoyopadhyay (1986) conducted experiment on 4.7 ha of salt affected community land in Gurgaon district of Haryana with seven fast growing selected plant species namely, *Prosopis juliflora, Acacia nilotica, Eucalyptus hybrid (E. tereticornis), Dalbergia sissoo, Leucaena leucocephala, Azadirachta indica* and *Albizzia lebbek*. They reported that mortality rates, three months after planting, were high for *Acacia nilotica* (27%) followed by *E. tereticornis* (19%), whereas height after two year was best in *E. tereticornis* (3.04 m) and least in *A. indica* (0.5 m). The fuelwood plantations on such sites are thought to be economically feasible.

Jambulingam and Fernandes (1986) observed that the farmers in Tamil Nadu state integrate numerous species of multipurpose trees and shrubs on their farm lands in close association with agricultural crops and / or livestock. The dominant among them are *Borassus flabellifer, Tamarindus indica, Ceiba pentandra, Acacia leucophloea* and *Acacia nilotica*. These woody perennials are better able to cope with poor conditions. Their increasing integration on farm lands represents a strategy to minimise risk of crop failure. Some species (e.g. *Prosopis juliflora* and *Delonix elata*) are deliberately used to ameliorate infertile or saline soils in order to permit the growing of annual crops. In addition to producing fuelwood, charcoal, fruit and fodder and providing many service functions, these species require fewer inputs and give the farmer a choice of management options in the event of poor crop growth. Further they state that the productivity of these traditionally managed and little studied agroforestry systems can be improved considerably by scientific intervention. Chopra and Bhatia (1987) revealed that *Cassia sturtii* is a very fast growing, drought tolerant, evergreen and high fodder yielding shrub in the semi-arid region of Rajasthan. Tejwani (1987) concluded that in the semi-arid zone the farmer determines the trees which are to be planted on the farm and the commonly planted trees are *Eucalyptus hybrid, Casuarina equisetifolia, Borassus flabellifer, Phoenix sylvastus,* custard apple, mango, tamarind *(Tamarindus indica)*, neem *(Azadirachta indica)* and mahua *(Madhuca latifolia)*.

## Studies on Farm Boundary Trees

Prajapati *et al.* (1971) reported that yields of jowar (*Sorghum vulgare*) were adversely affected as far as 30 m away from the *Prosopis juliflora* tree (4 to 5 times the height of the trees) when grown on farm boundaries in semi-arid areas. In order to study the interference effect of tree roots on associated agricultural crops Singh and Dayal (1974) dug a 15 m x 1 m trench along the field boundary of a 15- year old plantation of *Dalbergia sissoo* and *Acacia arabica* planted on peripheral bunds of cotton and tobacco fields and evaluated the crop yields. The elimination of *Dalbergia sissoo* roots from the crop root zone increased the yield of cotton from 956 kg ha$^{-1}$ to 1056 kg ha$^{-1}$. The yield of tobacco increased from 1206 kg ha$^{-1}$ to 1859 kg ha$^{-1}$ by pruning the roots of *Acacia arabica.*

Srivastava and Narain (1980) studied the effect of a 20-year old *Eucalyptus tereticornis* tree line planted on the farm boundary on crops under rainfed semi-arid conditions. Serious reductions in the crop yields were noticed of upto 10 m in the case of green gram, upto 5 m in black gram and only 2 m in case of sorghum.

Dhillon *et al.* (1982) studied the effect of shading by Eucalyptus trees planted in E-W and N-S directions on yield of three crops grown on the N, S, E and W sides of the tree rows. The trees planted N-S direction caused less reduction in yield than those planted in E-W direction. The yield reduction was lowest in crops grown on the E aspect. The shade tolerance was highest and yield reduction lowest in potatoes followed by rice and wheat.

## Studies on Silvopastoral Systems

Singh *et al.* (1979) stated that the forest and forage resources of India fall short of requirements but could, however, be increased through Silvopastural systems where grasses are grown in the interspaces of fodder cum fuel trees in marginal and submarginal lands having low fertility. Relwani *et al.* (1982). recorded the highest yield of dry matter in 809 days with the *L. leucocephala* variety cunningham interplanted with hybrid napier NB.21 *(Penisetum purpureum italic # P. americanum).*

In studies initiated during 1980-81 *leucaena* was sown on 15th July and fodder crops were sown between the rows on 25th July (monsoon season) and 10th Oct. (winter season). Green forage yield from *leucaena* alone was 154 q $ha^{-1}$. In association with sorghum total yield was 354 q $ha^{-1}$. (205 q $ha^{-1}$ *leucaena* and 149 q ha-1 sorghum). *Leucaena* in association with pearl millet also gave a good forage yield i.e. 300 q $ha^{-1}$ (Gill *et al.*, 1982). In the semi-arid region at Jhansi, the performance of *Leucaena leucocephala* was evaluated in a silvopastoral combination by Gill and Patil (1983). They found that the tree was compatible with a number of forage production systems. Total biomass production in association with *leucaena* was invariably higher compared to monocrop systems. Successively, Gill and Patil (1984) conducted trials at Jhansi for identifying suitable cereal fodder crops for intercropping with *Leucaena leucocephala* variety K 8. They observed maximum fodder production (505 q $ha^{-1}$) with sorghum + *Leucaena leucocephala.* In other studies, the annual yield of green fodder was determined from *Leucaena leucocephala* and three sorghum species, alone or intercropped with three rows of sorghum between *leucaena* 1.0 m apart. Sorghum was grown as a summer/monsoon crop and was followed by Brassicas in winter in all treatments. Total fodder yield was highest with *leucaena* + *Sorghum sudanense / Brassicas.* Finally Gill and Patil (1985) concluded that *leucaena* is a potential fodder plant that can be intercropped with high yielding sorghum fodder crops under intensive well managed conditions. In a series of trials during 1980-83 at Jhansi, *Leucaena leucocephala, Sesbania sesban* and *Desmanthus virgatus* were grown alone or intercropped with hybrid napier grass or guinea grass (*Panicum maximum*). Hybrid napier varieties were also grown alone or with *Leucaena leucocephala, S. sesban* or *hamata*. In other trials *L. leucocephala* was intercropped with setaria grass. Fodder yields were highest when *L. leucocephala* was intercropped with hybrid napier IGFRI-3 or IGFRI-6 (Gill and Patil, 1985).

Mann and Muthana (1984) conducted silvopastoral studies on the establishment and growth of fodder tree species namely, *Acacia tortilis, Albizzia lebbek, Azadirachta indica, Holoptelia integrifolia* in the integrated land use pattern in semi-arid regions of Rajasthan. They found that percentage seedling survival and growth in height of *A. totilis* continued to be higher (93% and 2.50 m) than *Azadirachta indica* (91% and 2.48 m), *Holoptelia integrifolia* (70% and 1.48 m) and *Albizzia lebbek* (62% and 1.34 m) respectively. Seeding with *Dichanthium annulatum* and *Panicum antidotale* continued to affect significantly the seedling survival and growth in height of trees. Further they recorded the mean maximum yield of grass forage under each tree species. The highest yield of 22.4 q* $ha^{-1}$ was obtained under *Albizzia lebbek* followed by *Holoptelia integrifolia* (22.2 q $ha^{-1}$) and *Azadirachta indica* (21.5 q $ha^{-1}$) and least under *Acacia tortilis* (19.6 q $ha^{-1}$). [* q = 100 kg]

Gupta and Muthana (1985) and Shukla and Sangal (1986) reported agroforestry practices to be more economical than sole forestry or sole fodder cropping. Patil and Gill (1986) observed that the total green fodder yields of sevri (*Sesbania sesban*) grown in 1:4 ratio with teosinte (Zea) or sorghum (as fodder crops) were greater than those of the crops grown separately. Sevri can also be used as green manure, as fencing and wind breaks and for fuelwood.

Mohatkar (1987) obtained 80-150 t $ha^{-1}$ $year^{-1}$ fresh herbage yield from *Panicum maximum* intercropped in *Leucaena leucocephala* grown for wood production and of 30-40 t from each of the two species when *L. leucocephala* was grown for fodder production.

Pathak and Roy (1987) described the silvopastoral system on degraded land in arid / semiarid regions of India, which involves the production of forage grasses and legumes with multipurpose trees used initially under a cut and carry system but latter grazed *in situ*. Forage production is superior to that from natural grazing land and improved grass land and provided year round feeding by the provision of tree leaf fodder.

Rai and Suresh (1988) studied the shade tolerance of arable crops grown in association with scattered trees on farms. A fodder crop was intercropped with *Acacia leucophloea* in Periyar district of Tamil Nadu and CSH-1 grain sorghum intercropped with *Azadirachta indica* in Anna district. The sorghum showed increased height and dry matter production under the tree canopy compared with the open.

Shekhawat *et al.* (1988) planted three fodder crops between rows of *Acacia tortilis* in a 15 year old plantation at Jobner in June 1984 and separately as sole field crops and harvested in August 84. The crops were bajra *(Penisetum americanum)*, cowpea (*vigna unguiculata*) and guar (*Cymopsis tetragonaloba*). The trees were harvested for fuelwood in March, 84 by cutting at 45 cm height and leaving the stump for coppice cropping. They noted that fuelwood was not much different with or without fodder crops. Green fodder yields were reduced by about 50 per cent when intercropped, while total financial returns from intercropping systems were higher by Rs. 1756-2106 $ha^{-1}$ over fodder crops alone. Thus these results provided circumstantial evidence in favour of adopting agroforestry over that of sole forestry of sole fodder crops.

## Studies on Agri-Silvicultural/Silvi-Agricultural Systems

Venkateswarlu *et al.* (1981) conducted trials on red chalka soils at Hyderabad and found that castor, sorghum and pearl millet yields were improved when interplanted with *leucaena* which itself yielded considerable fodder. Similarly, Das and Reddy (1982) observed that the yields of four grains namely, pearl millet, sorghum, castorbeans and pigeonpeas were greater when crops were grown in paired rows with *Leucaena leucocephala* than when grown alone on the Deccan plateau at Hyderabad. Prasad and Verma (1983) studied the effect of *leucaena* alleys on associated crops of sorghum, sorghum and pigeonpea and sorghum and black gram. The results presented in table-4, indicate some reduction in yields of pigeonpea and black gram grown in association with *leucaena* compared to pure crop, but the additional biomass yield from *leucaena* makes the combination more paying.

Table 4 : Yield of *leucaena* and intercrop*

| Leucaena alone | *Leucaena* with intercrops | | Crops alone | |
|---|---|---|---|---|
| 3.75m x 25cm – 3700 | i) *leucaena* (3.75m x 25cm) | 1800 | pigeonpea | 1650 |
| 75cm x 25cm– 14000 | + pigeonpea | 1280 | sorghum | 370 |
| | + sorghum | 380 | | |
| | ii) *leucaena* (3.75m x 25cm) | 1400 | pigeonpea | 1690 |
| | + pigeonpea | 1370 | black gram | 350 |
| | + black gram | 208 | | |
| | iii) *leucaena* (3.75m x 25cm) | 1200 | sorghum | 660 |
| | + sorghum | 460 | | |

* Air dried biomass of *Leucaena* and grain yield of intercrops Kg $ha^{-1}$

Balasubramanian *et al.* (1984) conducted experiment in which *Leucaena leucocephala* was grown in paired rows with an interspace of 3 or 6 m between the two pairs of rows and was cut to a height of 50 or 100 cm. Sorghum, bajra (*Penisetum americanum*) and red gram (*Cajanus cajan*) were grown either in the interspaces or in pure stands. They observed that sorghum and bajra when grown both as pure stand and intercrops gave higher grain yields than seed yields of red gram and the reduction in their yields as intercrops was also smaller, while *Leucaena leucocephala* intercropped with red gram in 6 m wide interspaces and cut to a height of 100 cm gave high fodder yields and seed yields and also produced more biomass in terms of fuel branch, roots and stump weight and gave the highest net returns.

Gill and Patil (1985) at Jhansi, established *Leucaena leucocephala* and sugarcane in August, 1983 and February, 1984 respectively, as monocrops or were intercropped with 3 rows of sugarcane between *leucaena*. In June 84, *leucaena* was cut at 0-200 cm height. Yield of fuel and fodder was recorded in December 1984. Sugarcane was harvested in November, 1984. Maximum production of sugarcane and *leucaena* was obtained with monocrops. In the intercropped treatments fodder and fuel production increased with cutting height and was greatest in uncut plots. Sugarcane yield was reduced in intercropped treatments and was severely affected by increase in cutting height.

Patil (1986) demonstrated yield increment of 63 per cent for pearl millet grain and 19 per cent for straw when the crop was grown between 10 m strips of fast growing trees in semi-arid and arid locations. The trees also produce sizable amounts of forage and wood. Studies of *Sesbania grandiflora* have indicated positive gains in crop production, in addition to the forage and fuel benefits from the intercrops. Green manure from these trees showed increased cereal yields when grown in rotation with *leucaena*, millet yields were increased by 10 per cent over control.

Singh and Singh (1987) while summarising the work on alternate farming systems for the drylands of the semi-arid tropics of India, highlighted the effect of the addition of *leucaena* loppings to the soil in the improvement of sorghum grain yield (table 5).

Table 5 : Effect of *Leucaena leucocephala* planting, lopping addition and levels of fertiliser nitrogen on the grain yield of *rabi* sorghum (Solapur, India).

| Treatments | Grain yield ($10^2$ kg ha$^{-1}$) N levels | | | Total N uptake (kg ha$^{-1}$) N levels | | |
|---|---|---|---|---|---|---|
| | 0 | 25 | 50 | 0 | 25 | 50 |
| Control (without *leucaena* planting and addition of lopping from outside). | 6.2 | 12.1 | 16.8 | 16.6 | 22.3 | 45.0 |
| No planting of *leucaena* but addition of loppings from outside. | 10.7 | 15.8 | 17.4 | 26.33 | 43.2 | 48.0 |
| *Leucaena* planting without addition of loppings. | 4.7 | 10.8 | 12.7 | 13.5 | 26.3 | 33.8 |
| *Leucaena* planting with addition of loppings. | 9.3 | 13.9 | 16.6 | 24.3 | 36.4 | 44.5 |
| C.D. 0.05 | 1.89 | | | | | |

It is interesting to note that the addition of biomass from other sites in combination with fertiliser nitrogen increased the yield of sorghum. However, the combination does not indicate the quantity of lopping added in the experiment.

Rai and Suresh (1988) found that the growth and yield of sorghum, maize and cowpeas were depressed when alley cropped with *Casurina equisetifolia, Leucaena leucocephala* and *Eucalyptus tereticornis* tree species.

## CONCLUSIONS

The present situation of the agroforestry research work in semi- arid areas of India may be summarised as follows.

The agroforestry research work was started at 20 locations during 1983 and then expanded to 31 centres during the Seventh Five Year Plan in various agro-ecological zones.

The earlier results indicate that the tree species, *Cenchrus ciliaris, Stylosanthes hamata, S. scabra, Leucaena leucocephala, Hardwickia binata, Albizzia lebbek, Azadirachta indica, Dalbergia sissoo, Acacia tortilis, Dichrostachys glomerata, Colophospermum mopane, Borassus flabellifer, Tamarindus indica, Ceiba pentandra, Acacia leucophloea, Acacia nilotica, Prosopis juliflora, Delonix elata* and *Cassia sturtii* are suitable for growing either on degraded forest areas or growing on farmlands in close association with agricultural crops. *Eucalyptus* hybrid (*E. tereticornis*) as well as *Prosopis juliflora* and *Delonix elata* are suitable tree species for saline soils/salt affected areas of the semi-arid region.

In the semi-arid situation the chances of failure of agricultural crops once or twice in a span of five years is a common phenomenon due to erratic precipitation. Under such situations growing of woody perennials on farm lands certainly represents a strategy to minimise risk of crop failure because of their better ability to cope with poor conditions. These tree species produce timber, fuelwood, charcoal, fruit and many other service functions which will pay the farmer in the event of crop failure and support to his economic condition. The tree species like *Prosopis juliflora* and *Delonix elata* can be used to ameliorate infertile or saline soils in order to permit the growing of annual crops.

The tree species *Dalbergia sissoo, Acacia arabica, Eucalyptus tereticornis* and *Prosopis juliflora* have an adverse effect on the growth and yield of agricultural crops when grown in close association with the crops or on the field boundaries.

The results of silvopastoral studies prove that agroforestry practices are more economical than sole forestry or sole fodder cropping. *Leucaena leucocephala* seems compatible with fodder sorghum, pearl millet and hybrid napier grass. The total biomass production can be maximised by growing any of the forage crops as an intercrop in *Leucaena leucocephala* rows. *Sesbania sesban* also seems suitable for growing in combination with maize or sorghum for increasing fodder production. The sorghum as a fodder crop seems to be a shade tolerant and can be grown under the canopy of scattered trees of *Azadirachta indica* in the field. Similarly, crops like pearl millet (*Penisetum americanum*), cowpea (*Vigna unguiculata*) and guar (*Cymopsis tetragonaloba*) can be grown in old plantation of *Acacia tortilis* (cut at 45 cm height) for maximising the financial returns.

The results of intercropping studies shows that the yields of castor, pearl millet, pigeonpea and black gram may be increased by growing in paired rows with *Leucaena leucocephala* or with *leucaena* alleys than sole/pure cropping. Sometimes the yields may reduce, but the additional biomass and fuelwood may compensate for the reduction of yield and make the system more profitable. Millet yields can also be increased to the extent of 10 per cent by growing in rotation with *Luecaena* and green manuring with *Sesbania grandiflora*. Similarly the sorghum grain yield can be improved by addition of *Leucaena* loppings to the soil.

Presently, there is a need to increase food, fodder, fuelwood and timber production besides achieving environmental protection in the semi-arid region of the country. In my opinion agroforestry is the only solution to overcoming these problems. Though, research work on few indigenous and exotic species has been done, there is a need to work on number of the tree species which are already grown by the farmers. Agroforestry research should have, as a high priority, the identification of appropriate tree species for their assigned role in the system in a particular environment. The interaction of the trees and crops/grasses with respect to spacing, shade, rooting patterns, competition for plant nutrients and moisture, soil fertility and compatibility between various tree and crop / grass species needs to be studied along with the pest and disease problems in detail at each centre in the zone for a specific period prior to recommendation to the farmers.

## REFERENCES

Anonymous (1987) The state of forest report, Govt. of India, Ministry of Environment and Forest, Forest Survey of India, 25 Subhash Road, Dehra Dun.

Anonymous (1989) India, Country Report. xiii Commonwealth Forestry Conference (!7-30 Sept, 1989), Rotorua, New Zealand. Ministry of Environment and Forests, Govt. of India, New Delhi.

Balasubramanian, T.N., Robinson, J.G. and Ravikumar, V. (1984) Study on the intercropping of *Leucaena leucocephala* with annual crops under dryland condition. Madras Agricultural Journal. 71 (10) : 677-680.

Bose, R.K. and Bandoyopadhyay, S.K. (1986) Economics of energy plantation in alkali soils of Indian semi-arid regions. Biomass. 11 (1) : 51-60.

Chopra, D.P. and Bhatia, N. (1987) *Cassia sturtii* and ever-green fodder bush for western Rajasthan. Indian Farming. 37 (9) : 35.

Das, R.B. and Reddy, N.V. (1982) Intercropping of *Leucaena* with grain crops. Leucaena Research Reports. 3 : 23-24.

Dhillon, G.S., Singh, S., Dhillon, M.S. and Atwal, A.S. (1982) Developing agri-silvicultural practices, studies on shading effect of Eucalyptus on the yield of adjoining crops. Indian Journal of Ecology. 9 (2) : 228-236.

Gill, A.S. and Patil, B.D. (1983) Subabul *(Leucaena leucocephala)* for forage production in agriculture in India. Bulletin of Indian Grassland and Fodder Research institute, Jhansi, India.

Gill, A.S. and Patil, B.D. (1984) Crop association studies in *leucaena* under rainfed conditions. Leucaena Research Reports. 5 : 24-25.

Gill, A.S. and Patil, B.D. (1985) Intercropping studies with *leucaena* under an intensive fodder production programme. Leucaena Research Reports. 6 : 36-37.

Gill, A.S. and Patil, B.D. (1985) Subabul for forage production under intercropping. A new approach. Indian Farming. 35 (6) : 37- 40.

Gill, A.S. and Patil, B.D. (1985) Agroforestry studies of *leucaena* with sugarcane. Leucaena Research Reports. 6 : 35.

Gill, A.S., Patil, B.D. and Yadav, C.L. (1982) Intercropping studies in *leucaena*. Leucaena Research Reports. 3 : 30.

Gupta, J.B. and Muthana K.D. (1985) Effect of integrated moisture conservation technology on the early growth and establishment of *Acacia tortilis* in the Indian desert. Indian Forester. 111(7) : 477-485.

Jambulingam, R. and Fernandes, E.C.M. (1986) Multipurpose trees and shrubs on farmlands in Tamil Nadu State (India). Agroforestry Systems. 4(1) : 17-32.

Mann, H.S. and Muthana, K.D. (1984) Arid zone Forestry (with special reference to Indian Arid Zone). Central Arid zone Research Institute, Jodhpur. (CAZRI, Monograph No. 23) pp 42.

Mohatkar, L. (1987) Intercropping of Guinea grass with subabul. Seeds and Farms. 13 (1) : 24-26.

Pathak, P.S. and Roy, R.D. (1987) Silvopastoral system for forage production on wastelands. Indian Farming. 37 (8): 10-11.

Patil, B.D. (1986) Indian Grasslands and Fodder Research Institute, Jhansi. Agroforestry research in Semi-arid tropics. A report on the Working Group Meeting held at Patancheru, Andhra Pradesh, India, 5-6 Aug., 1985. Beldt, R.J. Van den, Patancheru, A.P., India. pp 59.

Patil, B.D. and Gill, A.S. (1986) Sevri, a multipurpose shrub. Indian Farming. 35 (10) : 24-25.

Prajapatii, M.C. *et al.* (1971) Effect of lateral development of *Prosopis juliflora* D.C. roots on agricultural crops. Annals Arid Zone. 10 (2&3) : 186-193.

Raddi, A.G. (1981) Integration of forage forestry in afforestation programme of Maharashtra State. Indian Journal of Range Management. 2 (1/2) : 81-86.

Rai, R.S.V. and Suresh, K.K. (1988) Agrosilvicultural studies optimum species combination. International Tree Crops Journal. 5 (1-2) : 1-8.

Relwani, L.L., Nakat, R.V. and Khandale, D.Y. (1982) Intercropping of four *Leucaena* cultivars with three grasses. Leucaena Research Reports. 3 : 41.

Singh, A. and Dayal, R. (1974) Peripheral trenching for eliminating root effect of forest trees bordering agricultural crop in ravine areas. Annual Report., Central Soil and Water Conservation Research and Training Institute, Dehra Dun, India.

Singh, D., Mannikar, N.D. and Srivas, N.C. (1979) Effect of silvipasture systems on changes in soil fertility. J. Indian Soc. Soil Sci. 27 (4) : 486-488.

Singh, G.B. (1987) Agroforestry in the Indian Sub-continent : Past, present and future. Agroforestry a decade of development. (edited by Steppler, H.A. and Nair, P.K.R.), Nairobi, Kenya, International Council for Research in Agroforestry (ICRAF), pp. 117-138.

Singh, R.P. and Singh, S.P. (1987) Alternate farming systems for drylands of semi-arid tropics of India. Paper presented in the National Symosium on Alternate Farming Systems, IARI, New Delhi, February, 1987 (unpublished).

Shekhawat, J.S., Sen, N.L. and Somani, L.L. (1988) Evaluation of Agroforestry Systems under Semi-arid conditions of Rajasthan. Indian Forester. 114 (2) : 98-101.

Shukla, N.K. and Sangal, S.K. (1986) Preliminary studies on strength properties of some exotic timbers. Indian Forester. 112 (5) : 459-465.

Srivastava, A.K. and Narain, P. (1980) Competition studies on tree crops association with *Eucalyptus tereticornis* under farm forestry at Kota. Annual Report., Central Soil and Water Conservation Research and Training Institute, Dehra Dun, India.

Tejwani, K.G. (1987) Agroforestry practices and research in India. Agroforestry realities, possibilities and potentials. (edited by Gholz, H.L., Dordrecht, Netherlands, Martinus, Nijhoft, 1987), pp 109-136.

Venkateswarlu, J., Reddy, N.V., Das, R.B. and Rao, U.M.B. (1981) Role of *Leucaena* in intercropping in dryland farming. Leucaena Research Reports. 2 : 30.

# Role of Biotechnology in Overcoming Plant Stresses

*Ajay Sharma; R. Ganesh; A. Joshi and M.M. Simlot*

Rajasthan College of Agriculture : Udaipur

## ABSTRACT

Biotechnologies have contributed to the preservation of germplasm (genetic diversity) through collection of microbial cultures, selection of new strains of microbes and novel genotypes of cultivated plants, amongst the major thrust areas of research. It all began with the studies on plant cell, tissue and organ culture, consisting of growing either of these on nutrient medium *in vitro*, thus raising undifferentiated masses of cells called callus or suspension culture in liquid medium, as a novel approach. Though a majority of the work has been done on regeneration of plantlets through micropropagation, in some cases, it was used for the recovery of virus-free plants. This technique has potential application in agriculture, viz., in the production of somatic hybrids, organelle and cytosol transfer, genetic trasformation and germplasm storage. Moreover, biotechnology may prove to be a useful tool in the isolation of biochemical mutants, perhaps with improved nutritional quality, resistance to disease and tolerance to stresses like those of water, salinity, heat-shocks, herbicides and those caused by the environment. Genetic changes may also be brought about by the use of DNA, DNA-virus and transposons.

## INTRODUCTION

Land, water and energy have become limiting resources in the further improvement in agricultural production due to population explosion, especially in developing countries. The solution, therefore, lies in producing germplasm with high yield and low water consumption. While the traditional method of selection and development of such varieties required long period of experimentation, it was falling behind the requirement of rapid development of such varieties (Chopra and Sharma, 1991). Besides the factors stated above, the fall in agricultural production is also due to losses that :- (a) increased soil developing salinity thus decreasing the cultivable land; (b) loss of crops due to pests and pathogens; (c) loss of crop due to low and heavy rainfall, and finally, (d) uncertain weather conditions. (Srivastava and Menon, 1987). The development of tissue culture and other biotechnologies in the last twenty years has opened up new vistas and an unlimited scope for rapid development of the varieties required for increasing agricultural production. It is hoped that the agriculture scientist, by making use of the newer techniques, will be able to produce crops which are tolerant to the following stresses :

## COMMON PLANT STRESSES

### 1. Salinity –

The adverse effects of saline soils especially in semi-arid areas are, (i) low yield and limitation to choice of crops with poor fodder quality, mainly due to presence of high quality of molybdenum and a low amount of zinc causing nutritional imbalances, leading to diseases among livestock, (ii) difficulty in construction work, and (iii) run off, in excess, of the soil and unexpected floods causing low infiltration. Hence, in such conditions only a few crops manage to survive. In Europe, salinity stress has been tackled through biotechnologies such as tissue culture, protoplast fusion and genetic engineering (Sasson, 1984). Thus, salt tolerant cell lines, especially of lucerne, have been developed by scientists at Texas A and M University, USA (Winicov, 1990). Studies on pea plants using Southern blots have been done for enzyme localization in the biosynthesis of proline by another team at the Texas University, USA (Daniel, 1990). Scientists working on ice-plant, tolerant to salt stress, report that this plant has the ability to compartmentalize sodium chloride, for instance, into salt glands on the leaf surface, including other physiological adaptations such as induction of

CAM photosynthesis that allows higher water- use efficiency. The ice-plant studies have also revealed a salt- induced decrease in DNA methylation leading to the identification of a number of unusual genes as that for hydroxyindole-o-methyl transferase, whose function is yet not known (Bohnert and Mullet, 1990).

## 2. Moisture and Heat Stress –

Moisture and temperature are the major factors affecting the net production of healthy plants. In dryland zones, stress due to both moisture and high temperature (heat) need to be overcome using biotechnology to obtain stress- tolerant crop varieties, resulting in the improvement of both quality and quantity of foodgrain. Certain cereals like wheat and rice require water-logged conditions. The challenge is to grow them under water deficient conditions. Though ecologically adapted plants like xerophytes and those that can exhibit anti- transpirant activity are well suited to arid and semi-arid zones, biotechnology has made possible the idea of designing important crops for biological and environmental stress conditions, exploring traits such as available water, deep-root systems, water-use efficiency and osmotic adjustments (Dastane, 1987). Some laboratories in the US have used restricted fragment length polymorphism (RFLP) markers to score these traits for studies on drought. They report that genes are rapidly induced at the trascriptional level by decreased turgor, which at the biochemical level, consisted of a putative ion channel and a thiol process (Opena *et al.* 1987). In India, high temperatures often cause damage to crops during periods of water stress. The physical properties of lipids and proteins in membranes are virtually affected during temperature stress. Hence, the use of anti-transpirants like abscissic acid (ABA), phenyl mercuric acetate (PMA), Farnesol, etc., are proving useful to protect plants against damage caused by high temperature or poor water retention conditions (Srivastava, 1988). There is another condition of low temperature that causes chilling injury to plants, especially in temperate regions and of course, during winter in tropical zones. When the standard temperature is $5^{o}$ to $7^{o}$ C, some plant injury does occur even if the stage of ground frost is not reached. Very little information is available in this aspect with regard to the application of biotechnology to fight out low temperature stress, though research on winter crops have been conducted in USA reporting on cryobehaviour of plant plasma membranes (Hunt and Wagner, 1990).

## 3. Stress Caused by Pathogens and Pests –

Plant diseases may be caused by virus, bacteria and fungi. Most plants are immune to most diesases and owe their complete resistance to such factors like mechanical barriers to infection or total physiological incompatibility. Within a given species, there may be variation between individuals in susceptibility to potential pathogens (Burnet, 1991). In fact, new research comprising the use of biotechnologies like cell and organ culture, gene recombination, etc., are generating hope for emergence of bio-industries producting disease-free, genetically engineered plants. Since 1983, when life scientists originally developed the capability of plant trasformation in petunia and tobacco, more than a dozen of vegetables and commercial crops have been trasformed in many laboratories around the world (Sasson, 1984). Plants provide food not only to humans and parasitic microbes but also to herbivorous animals including insect, nematode and mollusc pests that reduce the crop yield. Hence, resistance to pests is another corollary to stress caused by parasitic microbes. The idea of using disease organisms to control pests is possible only by applying biotechnology to combat pathogenesis and pestilence that hinder productivity (Sasson, 1984). For instance, genetic engineers will have conferred commercially desirable traits, viz., resistance to pests and pathogens on many major crops like rice, corn, wheat, sorghum, cotton and alfa-alfa (Lamb, 1990). Plants have already been made resistant to some insects and viruses, giving the farmer manyfold benefits. For example, research scientists at the US have genetically engineered cotton to resist both caterpillars (pink boll-worm) and beetles. Thus, the effect on cotton production in the coming decade will be dramatic, putting behind the conventional agricultural practices (Goodin, 1988).

## 4. Resistance to Herbicides and Other Chemicals –

Genetic engineering allows development of herbicide-resistant crop varieties thus helping plant breeders with rapid programmes, the resulting variation thus proving to be environmentally friendly. These herbicide resistant crops will provide farmers not only an increase in productivity but also give important opportunities for new crops to be cultivated in rotation for effective weed control. For instance, several crops have been genetically engineered, in the US, for resistance towards round-up herbicide, glyphosate, thereby resulting in substantially lower cost inputs for weed control (Koundal *et al.* 1989). Of late, some of the approaches with regard to alleviating chemical pollutants from the environment of plants, in particular, which will enhance crop productivity tremendously, are photosynthetic improvers, antitranspirants, bio-fertilizers, bioinsecticides, bioherbicides that are all environmentally friendly - and of course, plant biotechniques such as tissue culture, protoplast fusion and genetic recombination, that can be very well commercialized in India (Natesh *et al.* 1985; Mulky *et al.* 1988).

## 5 Losses Due to Adverse Weather Conditions –

India being a land of many climates and varieties of soils, it affords scope for much diversity in agriculture. However, it is difficult to grow any crop in any region for that matter, due to genetic and adaptational differences. The losses due to such conditions include (a) poor chances of crop survival due to abnormal growth and development in deficiency/excess; (b) low productivity, on the whole, and (c) decreased resistance to the various stresses, to mention the major ones (Koteswaran and Raman, 1987). Here in lies the importance of germplasm preservation which has been made possible by recent advances in molecular biology, ranging from biosphere reserves to the creation of genebanks/DNA libraries (Swaminathan, 1988). Owing to man-made changes in the environment, germplasm resources in plants have been endangered. Deforestation, destruction of habitat and wide range pollution are noted examples of the intervention by man (Schmeiderman, 1987). There are two ways to store germplasm, viz., (i) slow growth, *in vitro,* and (ii) cryopreservation. In most cases, cryopreservation is the technique employed, though survival of such cells is about 50% when compared to the tissue meristems *in vivo* whose growth and regeneration ranges from 0-100% (Kartha, 1984). Reliable techniques in cryopreservation are being tried out, for example, rapid freezing by immersing cells directly into liquid nitrogen in the presence of protectants such as glycerol or dimethyl suloxide, etc. (Ahuja, 1986). Nevertheless, the immediate application of biotechnology to plants is in rapid and reliable clonal propagation of selected genotypes (Withers, 1985). Thus, germplasm collection of crops, their wild relatives and unrelated plants are essential if biotechnology is to be applied to find solution to problems posed by adverse weather conditions (Mantell and Smith, 1983).

## BIOTECHNOLOGIES APPLIED TO OVERCOME STRESS

The application of tissue culture to crop improvement has become quite popular in the Indian research scenario towards clonal propagation of elite genotypes of a variety of crop plants. *In vitro* plant propagation from apical meristems, for instance, has lead to producing disease-free germplasm, useful for pathogen eradication from any infected plant material (Jagannathan, 1983). Thus, the following *in vitro* techniques have proved the worth of biotechnology in creating genetic changes thereby modifying plantlets with new genotypic character and also development of resistance to both physiological and environmental stress conditions :

### (1) Hybridization

A standard technique to create variability in plants suffers from a serious limitation of incompatibility which restrict its scope to hybridize between closely related parents only. Using this technology, embryo and scutellar tissues can be dissected out at an early stage and cultured, thus obtaining hybrid plants. Following this technique, hybrids have been developed for mustard, jute, wheat, etc. (Mascarenhas *et al.* 1982). In India, virus-resistant papaya plants have been produced through embryo culture (Pandey and Rajeevan, 1983) and hybrid cotton (Joshi and Johri, 1972). As reported earlier, meristem culture has proved useful in obtaining virus-free plants especially in sugar cane and banana (Hendre *et al.* 1975; Nadgauda *et al.* 1982) and other major crop species.

Somaclonal variation through embryo culture has become very useful as enormous variability is exhibited amongst somaclones produced. Various resistant varieties against salt tolerance, heat shocks and resistance to certain pathogens have been made possible recently in potato, sugar cane, sugar beet and some vegetables (Mujeeb-Kazi and Stich, 1989). Somatic hybridization overcomes the limitations that occur in sexual hybridization and opens up the possibility of producing nuclear hybrids or even cytoplasmic hybrids (cybrids) between any two species, thereby involving fusion of protoplasts from the two parents. For example, in Japan, tomato and potato (Pomato) hybrids have been produced on commercial scale. Other crops like wheat and rye hybrids (triticale) and of turnip and cabbage (raphanobrassica) have been obtained (Huang, 1988).

### (2) Haploid plant production

Haploids are of great value as inbred lines can be directly obtained by chromosome doubling. Simple seggregation pattern in haploids make the selection of the desired type easy. Success in the production of pollen-derived plants (anther culture) has been established for a number of crop plants. Tremendous work has been done on rice in the Asian continent (Phillipines, IRRI) (Mori *et al.* 1985).

Using standard procedures, it has been possible to obtain from stable and desirable recombinants, potentially desirable $F_1$ hybrids, for example CO-40 is a good tillering and compact plant type in rice. For an early maturing cultivar in wheat, the genotypes of $R_1$ and $R_4$ have been selected by breeders, as they possess the desired attributes and bred true (Mascarenhas and Jagannathan 1980). Anther culture studies on certain legume crops have been successful in India, on a laboratory scale (Mohan Ram *et al.* 1982). A method has been standardized for haploid plant regeneration from anthers of *Annona* (custard apple) (Nair *et al.*, 1983). Anthers have also been used from cocoanut palm (Sarin and Bagga, 1988).

## 3. Selection of Mutants

Cell culture technique has several advantages over conventional breeding for inducing mutation in that the desired mutant is produced in the first generation itself and in large populations. Therefore, the mutated cells do not have to undergo the constraints of haplotic and diplotic seggregation. Contrarily, they ensure high recovery. Addition of selective chemicals to tolerate salt and water stress in certain desert plants, for instance, have resulted in creating mutation *in vitro.* Thus, mutants resistant to salinity stress or to herbicides and other toxic substances have been isolated in some crop plants (Chopra *et al.* 1989).

Rhizobium plant interaction has been the subject of intensive research in biotechnology. Infection of legume roots and fixation of nitrogen by rhizobia has been analysed in great detail at the genetic and molecular level. Rhizobia characteristically carry very large plasmids and these quite often carry the *nif* gene cluster. Long (1986) reported that there are two nod gene clusters, ABCD & EFGH, because of its resemblance to *Agrobacterium* spp. In addition to this, rhizobia carry similar viral genes which promote the transfer of T-DNA into dicots. Plasmids of Rhizobia enhance the rate of nitrogen assimilation (Mehta *et al,* 1989; Das, 1991).

## 4. Gene-transfer and Genetic Recombination Technology

The most direct approach of transforming a plant cell genetically is to introduce isolated genes (allien DNA) into a recipient cell. The general scheme of transformation of a higher plant cell necessitates the construction of a suitable carrier or a vehicle carrying the desired genes and ensure stable incorporation of these into the host cell. Amongst the potential carriers known today is the Ti-plasmid of *Agrobacterium tumefaciens.* This plasmid is now being genetically engineered for its tumour forming properties. Thus, using this plasmid, genes responsible for various resistances i.e. to antibiotics and pathogenic microbes, have been trasferred from bacteria to plant cells (Notani, 1988). Other methods like *biolistics* have been developed to shoot allien genes into the nucleus of the host cells and get them organized in the native genome to form a new genome resistant or tolerant to the stress in question (Mendel, 1990). The recombinant DNA technology has been speculated for creation of novel genotypes that can be (i) tolerant or resistant to diseases, pests, frost injury, salinity, toxicity of herbicides and pesticides; (ii) make efficient fertilizer utilization with minimum loss; (iii) photosynthetically efficient; (iv) fix atmospheric nitrogen, considering non- leguminous plants too, and (v) improve seed nutritional quality in terms of their storage protein (Mantell *et al.*, 1985). Of late, the transfer of uptake hydrogenase (*hup*) genes to Hup-strains of rhizobia, in the context of improving nitrogen fixation by legume crops, reward the plant system in cutting down energy loss during the process (Shende and Apte, 1982). A team of Indian scientists at the H.A.U., Hisar (Haryana) and IARI, New Delhi are actively engaged in the genetic analysis of uptake hydrogenase by techniques of molecular genetics used in the construction of improved strains by gene transfer (Garg *et al*, 1990).

Improvement in crop plants with regard to physiologically induced stresses has allowed genetic engineering to introduce transgenic plants that call for routine and efficient gene trasfer into any desired variety of any given species, changing the genome only by the addition of one defined gene (Gasser and Fraley, 1989). Hence, research is going on at a rapid pace to solve the problem of integrative genetic trasformation into zygotic proembryos, predominantly (Potrykus 1990 a, b).

The futurology of such research leads scientists to protecting intellectual property in plants through innovations in biotechnology, in the final analysis (Duesing and Raeber, 1989).

## REFERENCES

Ahuja, M.R. (1986) (Perspectives of Plant Biotechnology), Current Science, 55 (5) 217.

Bohnert, H. and J. Mullet (1990) *Ibid.* (p. 435). (*Ed. J. Mullet on "Designing crops for Resistance to Environmental stress").

Burnet, L. (1991) Exercises In Applied Genetics, Cambridge University Press (UK) Chap. 4, 47 (1991).

Chopra, V.L. and Sharma R.P. (1990) Current Science, 60 (9 & 10), 543.

Chopra, V.L., Narasimhulu, S.B., Kirti, P.B., Prakash, S. and Anuradha, G. (1989). Review of Advances In Plant Biotechnology (1985-1988) (Eds.) A. Mujeeb-Kazi and L.A. Stitch, IRRI Publ., 229.

Daniel, H. and Wang, L. (1990) *Ibid* (p. 437).

Dastane, N.G. (1987) Handbook of Agriculture, ICAR Publ., N. Delhi, Chap. 6, 158.

Das, H.K., (1991) Current Science, Vol. 60, (9 & 10) 551.

Duesing, J. and Raeber, J. (1989) Ciba-Giegy Journal. 3, 19.

Garg, N., Garg, R.P. and Nainawatee H.S. (1990) Indian J. Exptal. Biol. 28, 957.

Gasser, C.S. and Fraley R.T. (1989) Science, Vol. 224, p. 1293.

Goodin, J. (1988) Discoverer, Agriculture In 2001 A.D. (Science Magazine, USA). 9 (11) 60 (1988).

Hendre, R.R., Mascarenhas, A.F., Nadgir, A.L., Pathak, M. and Jagannathan, V. (1975) Indian Phytopathology 28, 175.

Huang, Cheng-Hua (1988) (Ed.) (Cell and Tissue Culture in Field Crop Improvement) Proc. Sem. Cell and Tis. Cult. Field Crop. Impr. Japan., F.F.T.C. Book Series (38).

Hunt, A. and Wagner G. (1990) Agricultural Biotechnology : News and Inf. 2 (4) 438.

Jagannathan, V. (1983) Present Status of Plant Tissue Culture in India)-Proc. Natl. Sem. Plant Tis. Cult. ICAR Publ., N. Delhi, Chap. 1, p. 3.

Joshi, P.C. and Johri, B.M. (1972) Phytomorphology. Vol. 22, 195.

Kartha, K.K. (1984) Cell Culture and Somatic Cell Genetics of Plants. Vol. I (Ed). Vasil, I.K., Academic Press, USA. 621.

Koteswaran, P. and Raman, C.R.V. (1987) Handbook of Agriculture, ICAR Publ, N. Delhi. Chap. 1, p. 1.

Koundal, K.R., Santha, I.M., Mehta, S.L. and Malik, V.S. (1989) Recent Advances In Plant Biochemistry, Chap. 11, 453.

Lamb, C. (1990) Agricultural Biotechnology : News and Inf. 2 (4) 443 (1990).

Mantell, S.H. and Smith, H. (1983) Plant Biotechnology, Cambridge University Press (UK).

Mantell, S.H., Matthews, J.A. and McKee, R.A. (1985) Principles of Plant Biotechnology. : An Intr. to Genet. Engg. Oxford : Blackwell Scientific Publ. (UK), 34.

Mascarenhas, A.F. and Jagannathan, V. (1980) IInd Natl. Sem. Genet.

Wheat Impr. H.A.U. Hisar, p. 1.

Mascarenhas, A.F., Hazara, S., Potdar, V., Kulkarni, U. and P.K. Gupta, (1982) Proc. Vth Intl. Cong. Plant Tis. & Cell Cult. Tokyo, Japan. 719

Mehta, S.L., Lodha, M.L. and Sane, P.V. (1989) Recent Advances in Plant Biochemistry. ICAR Publ., N. Delhi. Chap. 11, 456.

Mendel, R.A. (1990) Agricultural Biotechnology. News and Inf. Vol. 2 (5), 643.

Mohan Ram, H.Y., Mehta, U., Rao, I., Ramanuja, V. and Narasinham, M. (1982) Vth Intl. Cong. Plant Tis. Cult. & Cell Cult. Tokyo, Japan. 541.

Mori, K. Huang, C.R., Wu, Y.H. and Chen, C.C. (1985) Rice Genetics, Proc. Intl. Rice Genet. Sym. IRRI Publ., Phillipines, p. 753.

Mujeeb- Kazi, A. and Stich, L.A. (1989) (Eds.) Review of Advances in Plant Biotechnology (1985-1988) IRRI Publ., Phillipines.

Mulky, M.J., Srivastava, H.C. and Vatsya, B. (1988) Industrial Research, Oxford & IBH Publ. 50.

Nair, S., Gupta, P.K. and Mascarenhas, A.F. (1983) Plant Cell. Reptr. Vol. 3, p. 198.

Natesh, S., Chopra, V.L. and Ramachandran, R.S. (1985). Biotechnology in Agriculture, Oxford University Press.

Nadgauda, R.S., Khuspe, S.S. and Mascarenhas, A.F. (1982) Vth Ann. Sym. Plantation Crops. Abstract No. 23, CPCRI, Kasargod (AP) India, p. 15.

Notani, N.K. (1988) Industrial Research (BARC, Bombay). Crop Impr. Genet. Manipul., Chap. 3, p. 50 (1988).

Opena, T.R., Kuo, C.G. and Yoon, J.Y. (1987) Vegetable Production in Asia. (ASPAC), FFTC Book Series (36) Ed. Huang, Cheng Hua. 88.

Pandey, R.M. and Rajeevan, M.S. (1983) Proc. Natl. Sem. Plant Tis. Cult. ICAR Publ. N. Delhi, Chap. 5, 33.

Potrykus, I. (1990) Biotechnology, p. 535.

Potrykus, I. (1990) Physiologia Plantarum, 78 (1), (1990).

Sarin, N.B. and Bagga, S. (1988) Proc. Natl. Sem. Plant Tiss. Cult. ICAR Publ., N. Delhi. Chap. 19, 132.

Sasson, A. (1984) Biotechnologies : Challenges and Promises, Unesco Publ., p. 291.

Sasson, A. (1984) Biotechnologies : Challenges and Promises, Unesco. Publ. 239.

Schneiderman, H.A. (1987) Agricultural Engineering News. Monsanto Company, St. Louis, (USA) 28.

Shende, S.T. and Apte, R. (1982) Proc. Natl. Symp. Biol. Nitrogen Fix. p. 532.

Srivastava, H.C. (1988) Industrial Research (Ed). Mulky, M.J. *et al.* Oxford & IBH Publishing Co. Pvt. Ltd., N. Delhi, 60.

Srivastava, H.C. and Menon, K.K.G. (1987) Crop Productivity (Ed.). M.C. Srivastava, Oxford & IBH Publ. Co. Pvt. Ltd., N. Delhi.

Swaminathan, M.S. (1988) Industrial Research (Eds). Mulky, M.J. *et al.* Oxford & IBH Publ. N. Delhi. 22. Winicov, S. (1990) Agricultural Biotechnology : News and Information, 2 (4), 436.

Withers, L.A. (1985) In Vitro Techniques - Propagation and Long- term Storage (Ed.). A Schafer - Meunhr, Junk Publ., Dordrescht (Germany) 137.

Dryland Resources & Technology, 7 (1992)
Editors : L.L. SOMANI, R.L. SHYAMPURA
& M.C. MUNDRA


# Management of Alkali Soils in the Process of Reclamation

*L.L. Somani, K.L. Totawat,*
*P.V. Dave, D.L. Lavti & K.C. Laddha*[*]
[*] Rajasthan College of Agriculture
Udaipur

## ABSTRACT

Once the alkalinity of sodic soils is partially reduced to the level critical for survival of crop plants, further amelioration could be achieved through judicious adoption of agronomical and cultural practices. During the early stages of reclamation, special care has to be taken with regard to preparation of the field, selection of alkali tolerent crops and their tolerant varieties, proper plant population, spacing, age of seedlings and suitable time of planting and transplanting. Poor plant population can be taken care by higher seed rate/larger number of seedlings per hill besides close spacing. To some extent alkali tolerance could be inducted by treating seeds/seedlings with salts before planting. Sodicity of irrigation water could be controlled by mixing gypsum or neutralizing it by sulphuric acid or by blending water supplies. These agromeliorative practices encourage plant growth, the vary growth of plants leads to increased addition of root residue and $CO_2$ evolution which forms carbonic acid to neutralize the alkalinity.

## INTRODUCTION

The problem of soil alkalinity is one of the serious factors restricting crop production particularly under irrigated conditions in the developing countries of semi-arid and arid regions. Incorporation of amendments to replace adsorbed sodium from soil exchange complex is by far the most common method in restoring the productivity of such soils. The use of amendments is getting more and more costlier day by day because of energy crisis. Moreover, use of amendments will lead to partial success unless suitable agronomic and cultural practices are adopted. On the other hand it is being increasingly realised that dependence on chemical amendment for reclamation of alkali soils could be considerably reduced once these soils are partially reclaimed to the point critical for the survival of plants, further reclamation could be achieved by making judicious use of agronomic and cultural practices, referred to as agromelioration.

## AGROMELIORATIVE PRACTICES

### Mulching

Mulches not only reduce water losses by evaporation but also reduce salt accumulation on the surface of the soil by disrupting capillary continuity. Evaporation is known to decrease the effectiveness of reclamation measures by encouraging the ascent of initially downwardly leached sodium (Makrimyuk, 1958). An increase in the population of *Azotobacter*, *Nitrosomonas* and *Nitrobacter* is also associated with the application of mulch (Bajpai *et al.* 1976). Hence surface application of crop residue and placement of sand or gravel layers on the soil surface or slightly beneath is worth practicing.

Khosla and Yadav (1976) in a field experiment on a sodic soil, observed that addition of 20 to 30 t/ha of rice husk in upper 10 to 20 cm soil resulted in a significant improvement in the filteration rate and increased paddy yields. Bandyopadhya and Sen (1977) observed that the leaching efficiency of ponded water during the monsoon was improved by surface mulching with rice husk, apparantly due to reduced evaporation and less upward movement of water under mulch.

Use of sand in the form of mulch (3-8 cm thick) on the surface of sodic soil brought about appreciable improvement in the water permeability of the soil. However, mixing of sand did not result in substantial improvement (Acharya and Abrol, 1976). Filling the auger holes with porous material like sand accelerated leaching of soluble salts from saline-alkali soils (Chahal, 1962). Anjaneyulu (1972) obtained maximum response to vertical mulches (10 cm diameter to a depth of 2.5 m) of sand and organic material at the rate of 100 per hectare. Addition of gypsum to the combination further improved the response. Neverthless, the effectiveness of these materials need to be studied in greater detail in different types of salt affected soils and for soils irrigated with saline waters.

## Tillage

It is known that salts tend to accumulate closer to the soil surface in unirrigated drylands and to move downward with reclamation and irrigation. Hence, tillage to loosen the dense subsoil prior to start of irrigation will speedup the movement of soluble salts deeper in the soil upon irrigation. Deep ploughing and subsoiling are considered helpful in improving physical condition of the soil and in loosening up of the dense subsoil layer. Russian scientists particularly emphasise the benefits of deep ploughing. They recommend either ploughing to a depth of 30- 40 cm or else subsoiling. When the soil is heavy, with a compacted horizon underlying the ploughed layer, it must be loosened to a depth of 40-45 cm before irrigation (Rabochev, 1954).

Cairns (1969) reported greater removal of salts due to deep ploughing of the soil to a depth of 50 cm as compared to ordinary ploughing (10 cm). The yield of alfalfa-brome grass crop was doubled. Prettenhoffer(1964) found subsoiling to have greater and lasting effect on alkali soils. Subsoil loosening was found to be better than deep ploughing (Prettanhoffer, 1974).

Dundanov (1954) in his field experiments on highly saline chestnut soil with the water table at 6 m, found application of 840 kg/ha, $CaSO_4$ followed by ploughing to 30-35 cm (inverting the saline horizon) and irrigation to increase the yield of winter wheat by 40% and decrease the exchangeable Na from 9.67 to 1.62 me and the salt content from 155.8 to 56.0 me/100 g of soil. Breaking of compacted subsoil layer may not be advisable under conditions of high water table. Novikova (1961) reported 10-20% lower yields of wheat on meadow solonetz due to deep ploughing (50 cm) than with ordinary ploughing(18-20 cm). Intensive upward capillary movement of solutions of salts in summer with destruction of solonetz horizon explains the decreased yield. Nadezhdin(1957) recommended a lowering of the water table and the application of gypsum in addition to deep ploughing for amelioration of sodic soils in the middle Trans-Ural region.

Prettenhoffer (1963) observed increased effciency of amendments with increasing depth of the loosened subsoil from 30 to 60 cm. Subsoil loosening led to leaching of soluble salts beyond 60 cm depth in the profile, as compared with 40 cm or less without subsoiling. He further reported that the effect of subsoiling to be larger and lasts longer on slightly alkali soil. He also developed a cultivator with three blades able to cope with the hard resistance of alkali soils. According to Sipos (1975) the deep loosening increases the soil's pore volume and parallel with this the water reserves of the soil layer loosened to 60 cm depth on the average of three years was found to be more compared to the soil layer in natural physical state. Even on soils having been ameliorated chemically earlier, it was found necessary to make the physical state of 20-60 cm soil layer more favourable by deep loosening carried out periodically. Deep ploughing and subsoiling are thus very effective in alkali soils containing a compact B-horizon which obstructs water percolation and root penetration.

Alkali amelioration with gypsum is effective but the process is costly because the gypsum has to be brought in and spread. Satisfactory "self amelioration" can be achieved by ploughing to a depth of 60 cm, turning onto the surface B-horizon containing $CaCO_3$ and $CaSO_4$. Viewing from morphological viewpoint deep ploughing involves alteration of morphological features to produce a normal profile. This measure gives immediate results in improvement of physical conditions and higher yield next season. Pak (1957) lists the advantages of deep ploughing for cultivating solonetz soils. The dilution of the solonetzic layer with the more normal A horizon and the gypsiferous or calcareous C-horizon results in stimulation of biological process, especially $CO_2$ production and increased solubilization of lime. The benefits from one overturning of the soil last 15 years whereas the gypsum treatment has to be repeated at frequent intervals (Semenova- Zabrodina, 1952). Most solonetz like patches can be improved by deep ploughing to bring naturally occurring gypsum and carbonates to the surface (Sokolvoskii, 1956). Trials have shown that sodic clay pan soils of western North Dakota can be improved by deep ploughing. Ploughing 60-75 cm deep crumbles the dense subsoil layer and brings up natural gypsum from below the pan. Density is reduced, permeability and water holding capacity are increased and crop production is benefitted (Sandoval, 1978).

According to Botov (1959) the most suitable method for the radical improvement of solonetz soil seems to be the redistribution of genetic soil horizons in such a way that humus horizon, which is comparatively more fertile, is kept at the surface but the solonetz and gypsious horizons change place. In devising equipment for 2 or 3 stage plough, it is important to consider the soil morphology, salt concentration and distribution, occurrence of gypsum and lime, exchangeable sodium, and depth to the water table (Antipov-Karataev, *et al.* 1957). Implements have been developed for such a soil menipulation. Balyabo *et al.* (1956) found 3-layer ploughing to 50 cm in which the more fertile surface horizon is loosened *in situ* and the middle and gypseous horizons are mixed, was superior to ordinary ploughing to 50 cm. However, if the subsoil is devoid of gypsum, mere subsoiling or deep ploughing of alkali soils may not be effective. Balyabo (1957) also found three layer ploughing to be superior over other methods of ploughing in the reclamation of a steppe-solonetz soils. Moreover, if the subsoil is more alkali, deep ploughing with turning over of soil will aggravate the existing problem at the surface. This is the reason why Agarwal *et al.* (1979) failed to get any response to deep ploughing (45 cm) in salt affected alluvial tracts of India. Deep ploughing of sodic soil led to further deteriorate the already deteriorated soil physical condition (Bhumbla, 1972).

Ploughing to 30 cm without application of gypsum was less effective on non-gypseous solonetz soil than very deep ploughing (65 cm) but when gypsum was applied, ploughing to 30 cm was almost as effective as very deep ploughing. Very deep ploughing increased the soil moisture in the deep horizons of solonetzes and improved biochemical process occurring in the soil (Sambur, 1954). Three-layered ploughing together with application of gypsum is recommended for average solonchak-solonetzes, where A horizon is quite fertile and subsoil is devoid of gypseous layer (Zvereva, 1958). Novikova (1956) recommended ordinary ploughing to 18-20 cm and application of 3.5 to 4.0 t/ha of gypsum for meadow solonchak-solonetzes, which have very saline ground water with a high water-table and a high chloride content in the saline soil horizon. Meadow-steppe solonetzes in which the carbonate horizon occurs at 40-50 cm should be ploughed deeply so as to bring a 7-10 cm layer of carbonate/gypseous horizon to the surface, but when the carbonate/gypseous horizon occurs below 50 cm, ordinary ploughing and 3 t/ha of gypsum are recommended steppe-solonetzes on quaternary clays and with a low exchangeable Na content should be ploughed to 30 cm and 1.3 to 1.5 t/ha of gypsum be applied. Kertsch solonetzes should be ploughed deeply according to the depth of the natural gypsum layer, but when this depth is too large, ploughing to 30-40 cm and application 5 t/ha of gypsum are recommended.

The poor physical condition of alkali soil makes them to puddle when wet and develop a hard crust when dry. Therefore, use of heavy machinary should be avoided as it may further deteriorate their structure. Ploughing sodic soils at excessive moisture should be avoided as it destroys the aggregation and results in clod formation which ultimately reduces soil permeability. Deep ploughing with turning over of soil is no doubt beneficial in alkali areas having gypseous subsoil, yet care should be taken to avoid turning up the soil at advance stage of reclamation failing which salts will come up again. Although the practice of deep ploughing (with or without turning over) and subsoiling may be advantageous under some situations, there is need to undertake systematic investigations to evaluate different techniques of ploughing and subsoiling alone as well as in conjunction with organic and inorganic amendments inorder to elucidate suitable tillage technology for widely differing salt affected soils.

## Green Manuring

Green manuring is a practice of ploughing or turning in the soil of undecomposed green plant material for the purpose of improving soil productivity. Green manuring serves, on decomposition as source of readily available nutrients besides acting as solubilizing agent for calcium and neutralizing high pH of sodic soils. Consequent improvement in soil permeability and biological activity, help in regenerating the soil.

Amongst the several leguminous green manuring plants *dhaincha* has been found to be most efficacious, particularly in sodic soils. Dhawan et al. (1958) investigated the relative role of different green manure crops, viz., *dhaincha (Sesbania aculeata)*, guara (*Cyamopsis psoralioides*) and sannhemp (*Crotolaria juncea*) and reported that all these green manure crops contained sufficient calcium for replacing sodium from sodic soils. The most efficacious was *Dhaincha* which had the highest calcium content (34.2% Cao on ash basis) and the most acidic juice (pH 4.0). Talati (1947) and Kanwar (1962) also found *dhaincha* to be the best of the green manure crops because of its tolerance to sodicity, salts and water-logging, extensive root system, easy rotability, more acidic sap and more luxuriant growth than guara or sannhemp inspite of the fact that the N content of guara is higher than that of *dhaincha.* Uppal (1955) reported that the N content of oven dried stem and leaves of *dhaincha,* guara and sannhemp, one month after sowing were 2.86,3.18 and 2.01% respectively. Although, *dhaincha* contains less N than guara, but because of its high yield than guara, *dhaincha* has been preferred on the basis of its overall nitrogen content. Moreover, because of its extensive root growth, *dhaincha* improved the N status of soil much more than others. The increase in N content of soil adjacent to roots were 0.098, 0.080 and 0.077 per cent respectively with *dhaincha,* guara and sannhemp. *Dhaincha,* when transplanted, grew well in sodic soil with pH 9.5 but guara and sannhemp could not. The extensive root ramification of *dhaincha* improves the soil permeability, liberates $CO_2$ and organic acids which help to dissolve Ca from native $CaCO_3$ and thus help to reduce the ESP of sodic soils. Somani and Saxena (1981) found green manuring with *dhaincha* to be effective in improving the physical properties of sodic soils.

The mineralization of green manure, which is achieved generally uner flooded conditions, results in release of $NH_4$-N which is highly susceptible to volatilization in sodic soils. The extended period of decomposition following turning under of green manure crop may result in poor fertilization value. Best results are obtained by transplanting rice immediately after the turning of the green manure (Bhardwaj *et al.* 1977). In an incubation experiment, Venkat Krishnan (1980) observed extensive volatilizatior of N after 10 days of incorporation of green manure, implying thereby that it is not advisable to allow a decomposition period of more than 10 days between turning under of green manure crop and rice transplanting.

Tiwari *et al.* (1980) studied the direct and carryover effect of green manure on crop yield and nutrient availability in a partially reclaimed alkali soil and found thant N at the rate of 40 kg/ha with green manuring compared well with 120 Kg N/ha without green manure in respect of grain yield of rice. Dargan and Chillar (1980) reported 81.8 per cent increase in paddy yield without N application following green manuring as compared to fallow in a sodic soil.

Somani (1981) reported that the green manuring with *dhaincha* gave results as good as that of gypsum in improving the physical condition and reducing the ESP of a sodic soil. Yadav and Agarwal (1961) on the other hand, reported green manuring of *dhaincha* to be better than application of gypsum at the rate of 7.5 t/ha in a sodic soil with pH 10.2 and ESP 52. Somani and Saxena (1982) observed additive ameliorative effects when green manuring of *dhaincha* was used in conjunction with gypsum (Fig.1).

## Alkali Tolerant Crops

Crops such as wheat, barley, cowpea, pearl millet, etc. totally fail to grow under highly sodic conditions. There are, however, some grasses known as alkali halophytes (Joshi *et al.* 1981) which are known to grow well especially in the rainy season. This shows that it is not only the soil factor alone which leads to the failure of crops but it is also due to the sensitivity of the crops that they are unable to grow under high alkali conditions. The plants which grow well under alkali conditions have some specific characteristics as a result of which they resist alkali injury through the complex process involving alkali avoidance and tolerance. Besides the genetically inherited characteristics for alkali tolerance, physiological changes including chemical composition in the root, stem and leaves take place to some extent in the plant to survive

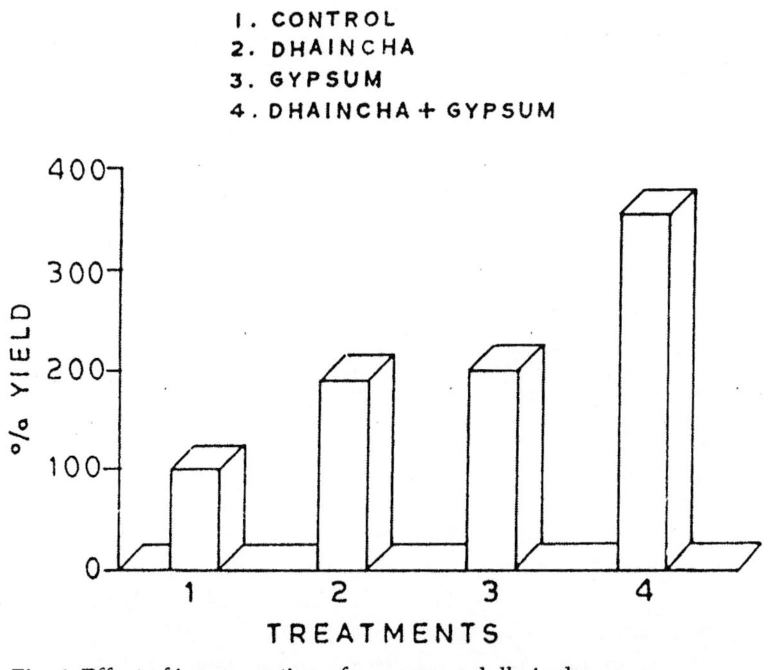


Fig. 1. Effect of incorporation of gypsum and dhaincha green manure on percentage yield of paddy in an alkali soil

under alkali stress conditions. Different crops have variable tolerance and capacity to produce satisfactory crop yields. Relative alkali tolerance limits of some of the important field crops are given in table 1.

Table 1 : Relative tolerance of crops to exchangeable sodium

| Tolerant | Semi-tolerant | Sensitive |
|---|---|---|
| Bermuda grass | Wheat | Cowpea |
| Para grass | Barley | Gram |
| Rice | Oats | Groundnut |
| Sugarbeet | Raya | Lantil |
| | Senji | Mash |
| | Berseem | Mung |
| | Sugarcane | Peas |
| | Millets | Maize |
| | Cotton | Cotton (at germination) |

Note : Crop yields are seriously affected if the ESP is more than about 55, 35 and 10 in respect of tolarant, semi-tolarant and sensitive crops, respectively, Tolerance in each column decreases from top to bottom.

Source : Abrol and Fireman (1977)

Rice is a well suited crop during reclamation of sodic soils. There is practically no reduction in rice yield upto an ESP of 55 and only a 20 per cent yield reduction at an ESP of 75 (Abrol and Bhumbla, 1979). The high tolerance of rice to alkali conditions is due to its rather superficial root system and its need for a layer of water on the soil during its growth. Moreover, the high pH of alkali soils, the other limiting factor for successful crop growth is reduced under continuously flooded conditions by the evolution of large quantities of $CO_2$ by microorganisms and plant roots and its accumulation due to restricted diffusion of gases in flooded soils (Sankaram, 1969, Ponnamperuma 1972).

Barley is another potential crop which can be grown in the initial stages of reclamation. A sharp decrease in the yield of barley has been observed when the pH of surface soil is higher than 9.4 (Milap Chand *et al.* 1977). The grain yield of wheat starts reducing beyond an ESP of 20. Fifty per cent yield reduction, however, takes place beyond an ESP of 50.

Among leguminous crops, *dhaincha (Sesbania aculeata)* is well known for its tolerance to high alkali conditions. The crop underwent nearly no reduction in green matter yield upto and ESP of 45, but at higher ESP, there was a reduction in yield. It is thus less tolerant than rice but more tolerant than wheat (Abrol and Bhumbla, 1979). Lentil and gram are sensitive crops. They gave poor yields even when the ESP of surface soil layer was less than 10. Cowpea is also sensitive to high alkali conditions (Dwivedi, 1979). It can be grown successfully only after a few years of reclamation.

Among oilseed crop raya (*Brassica juncea*) is the most promising crop for alkali conditions (Rana *et al.*, 1980). Kumar and Abrol (1979) investigated the alkali toerance of five grasses at different ESP. Their results (Fig.2) show bermuda grass and para grass to be highly tolerant while guinea grass almost failed at an ESP of about 25. Setaria and hybrid napier, on the other hand proved intermediate.

## Crop Rotation

Crops vary in their water requirement as well as the frequency with which they need water. Crops with long duration of evapotranspiration will enhance much more accumulation of Na on the exchange complex as compared to short duration crops in the absence of adequate leaching. Thus crops like rice, alfalfa, berseem requiring frequent irrigation helps the reclametion whereas crops like cotton result in increased accumulation of Na. A best crop rotation is one which not only provides good returns but also help to control or reduce the sodicity of the growing soil. Saline-alkali or alkali soils need introduction of crops having high water requirement (Bains, 1971).

Much of sodification in irrigation development throughout the world results from an inadequate water supply. Attempts are made to irrigate too much land in relation to the water supply. As a result of under irrigation there is little, if any, net downward movement of water through the root zone to satisfy the leaching

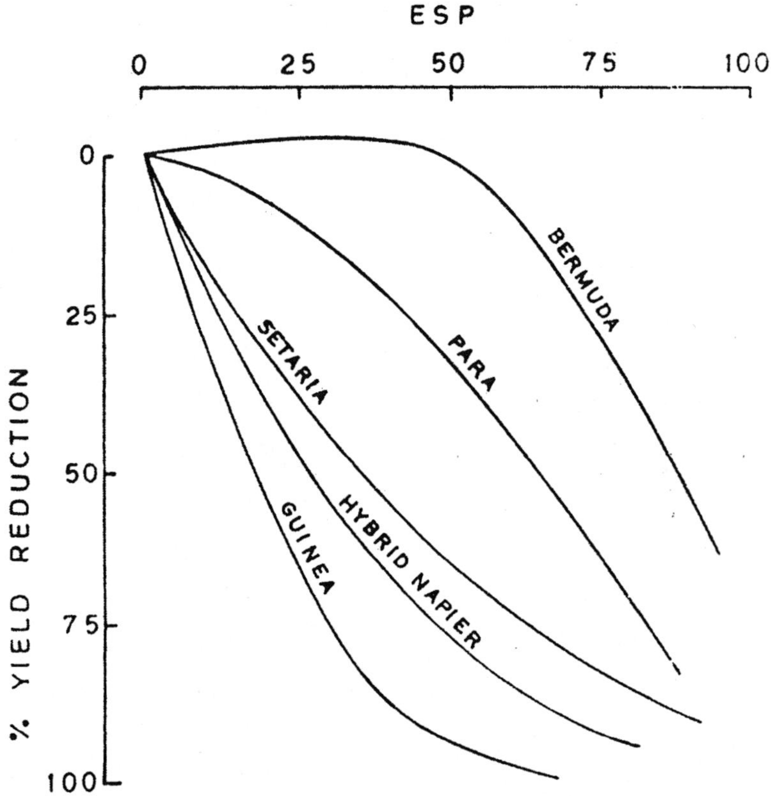


Fig. 2. Percent yield reduction of five grasses at different E.S.P. level

needs. The problem of underirrigation could be managed to a considerable extent by adopting following three measures :

(1) The total area irrigated may be reduced by an amount sufficient to provide enough water to meet the evapotranspiration and leaching requirements on the remaining area under the prevailing cropping pattern.

(2) Crops having a lower evapotranspiration requirement because their growth period is shorter or occurs when the evapotranspiration potential is lower may be grown under conditions of limited availability of irrigation water. Moreover, substitution of a crop having a low evapotranspiration requirement for one having a high requirement also reduce the LR, provided the alkali tolerance of the crops are equivalent.

(3) The cropping pattern may be altered in the direction of growing more alkali tolerant plants that have a lower leaching requirement thereby decreasing the depth of irrigation water required.

The growing of crops having low ET requirements and/or high sodium tolerance is a good alternative to compensate for the deficiency of water supply and soil permeability otherwise needed to achieve the LR. Owing to method or intensity of irrigation, cropping, season, and perhaps other factors, soil may tend to accumulate sodium during the growing of one crop and to dealkalise during the growing of another. Thus it is possible to achieve the LR over a period of time by manipulating the pattern of cropping. Because it is grown under flooded conditions, rice is particularly useful crop to include in a rotation for sodicity control, especially on soils having low permeability.

Since crops vary in the degree to which they can tolerate excessive salinity or alkali in the growing medium, the selection of first crop will depend on the nature and extent of salts left in the soil after initial reclamation measures including use of amendments and leaching etc. have been implemented. Some crops, while performing well, also contribute to the process of reclamation. Growing of a crop with extensive root system like rice often promotes a more favourable physical condition in alkali soils. Basu and Tegare (1943) tried several crops on mild alkali soils of Deccan canal area for two years. Crops like rice and berseem and *dhaincha* showed significant increase in exchangeable calcium of the soil. Exchangeable sodium was lowered in all the cases with berseem and *dhaincha* giving the highest figures. From increases in Ca/Na ratio, the order of superiority in ameliorating alkali character was found to be : berseem >rice >*dhaincha* >cotton >jowar >tobacco > wheat.

Yadav and Agarwal (1959) suggested that rice should form the first crop preferably after burial of *Sesbania* and should be followed by an alkali tolerant crop like barley. Satyanarayan (1953) suggested rice-rice-rice-pulse considerably pushes down salts enabling the cultivation of pulse in the second year. The second rice crop in the first year forms a temporary substitution in place of a pulse crop. Dhawan *et al.* (1958) reported leaching followed by the growing *Sesbania* and subsequent incorporation as green manure to be more effective for reducting alkalinity than was leaching-rice-gram rotation. Swank (*Panicum colonum*) followed by berseem was recommended for soils with pH 9.5. Sugarcane was the least effective.

The effect of four crop rotations viz., rice-wheat; rice-wheat-*dhaincha,* rice-berseem and rice-persian clover (Shaftal; *Trifolium resupinatum* Linn.) on crop growth and soil properties was evaluated during the reclamation of sodic soils (Sharma *et al.* 1983). The grain yield of subsequent rice crop was higher after rice-Persian clover and rice-berseem rotations than after rice-wheat rotation. Soil properties upto the depth of 30 cm showed maximum improvement after rice-berseem, followed by rice-Persian clover, rice-wheat-*dhaincha* and rice-wheat rotations. It was more profitable to raise Persian clover or berseem for fodder in winter than wheat. Normally, berssem is recommended as an alternative to wheat during winter. Experiments of Kumar *et al.* (1981) have shown that Persian clover, another promising fodder can also be included in place of berseem.

Johnson (1979) reported increasing salt affected area in Saskatchewan to be associated with summer fallowing. Fallowing encourage the upward movement of saline sodic groundwater and accumulation of Na on soil surface and on the exchange complex. Hence, as far possible, the practice of fallowing should be discouraged. However, if the ESP of soil is very high and good quality water is not available for irrigation, lands have to be kept fallow. Herein, lies the importance of the rainfall during monsoon season. The rainfall should therefore be conserved very efficiently so that leaching of sodium salts from the soils becomes very effective after incorporation of amendments. In long run, the allotment of water to the land and the capacities of the irrigation water conveyance and drainage system should be designed to permit near maximum cropping intensity.

Petrosian (1956) observed good growth of vine seedlings on soils shaded by irrigated lucerne due to decreased capillary rise of mineralized groundwater and to increased translocation of salts into deeper horizons. Bains *et al.* (1969) reported that the crop rotation of rice-barley or wheat-cowpea is highly effective in providing a continuous cover to the soil throughout the year and is able to stop salt accumulation in the surface soil in high water table areas. Asghar and Khan (1958) were successful in preventing damage to productivity by adopting suitable crop rotations in a soil where salts were rising towards the surface. When salts had risen up to 60 cm depth from the surface, a two year rotation with rice-cotton-senji was initially required to depress the salts below 80 cm depth. However, when the salts were below 180-240 cm depth, a four year rotation cotton-senji- sugarcane-wheat-rice was found adequate for keeping the salts depressed below that depth.

Kanwar (1969) suggested the possibility of introducing sugarbeet in crop rotations for saline-alkali soils of India. Mishra *et al.*, (1973) and Singh *et al.*, (1969) reported the preliminary results on the adaptability of this crop to saline-alkali conditions. Sugarbeet will not only yield good returns on saline alkali soils but also will increase the working days of sugar factories, which have seasonal work in India. Sugarcane-cum- sugarbeet factories have already been commissioned at some places, like the one at Beja in Tunisia (Bull, 1962), and it is possible to do the same thing in India or elsewhere. Another important factor in favour of sugarbeet introduction is that, unlike sugarcane, sugarbeet occupies the land for a comparatively shorter time and permits the cultivation of more crops on the same land during the year. Thus it can be fitted in the double- cropping scheme under irrigated conditions. Germination studies on sugarbeet and other crops on sodic soils revealed that sugarbeet can tolerate fairly high sodicities during germination as compared to sweet clover (senji) and berseem, the crops usually recommended for cultivation after rice on such soils (Kanwar and Singh, 1968). Sugarbeet yields were slightly affected at higher salt levels while yields of other crops such as sweet clover (senji), wheat and berseem are significantly decreased. Sugarbeet, therefore, seems to be very promising crop which could fit into crop rotations followed on saline-alkali soils. Totawat and Singh (1976) also reported that sugarbeet can be successfully grown as a winter crop in light textured soil with a saline sodic water of EC 4 mmhos/cm and SAR 22.

A number of workers have successfully tried different crop rotations under sodic environment. Some of the commonly used crop rotations for sodic soils are summarised in table 2. These crop rotations could be followed for the first 2-3 years and subsequently less tolerant crops like raya (*Brassica juncea*), sugarcane, gram, and even vegetables could be introduced.

Mere economic return should not form the basis of determining the superiority of one crop over the other but improvement in soil properties due to cropping should be given due consideration. Not only the efficiency of reclamation is slow but the economic returns are also poor in the absence of judicious selection of crops in crop rotation.

## Seed/Seedling Treatment

The plants which are able to maintain their chemo-osmotic potential in their cells under alkali stress have the capacity to withstand soil sodicity. One of the main causes for the poor germination or stand of the crop in alkali soils is the disruption in the permeability of roots due to which it fails to control balanced ion absorption.

Quadar *et al.* (1979) concluded experiments on effect of chemical treatment of wheat (cultivar, WG 357) seeds on its tolerance to sodicity. The seeds were treated with 3% solution of calcium

Table 2 : Suitable cropping sequences for amelioration of sodic soils

| Crop sequence | Desirable soil situation | Reference |
|---|---|---|
| Karnal Grass (*Diplachne fusca*)<br>Perennial or Para grass (*Brechiria mutica*) | Can be successfully grown on highly sodic soils (pH 10.7) particularly during initial 2-3 years after which normal crop sequences like Paddy-wheat-Sesbania, Sesbania-Suarcane-Cotton, etc. could be followed. They reduce soil pH. Most suited to conditions where use of traditional amendment is not practical or uneconomic. | Kumar *et al.*, 1980<br>Rana & Prakash, 1980<br>Rana *et al.*, 1980 |
| Paddy-Berseem/Barley<br>Paddy-Wheat-Sesbania | Remunitrative in the early phase of reclamation on highly alkali soil with pH above 10.5 | Agarwal *et al.*, 1979<br>Dargan, 1979 |
| Paddy-Wheat/Barley | Wheat perform better than barley in soils with pH upto 9.0 while barley perform better than wheat in soils with pH 9.0–9.3 or even upto pH 9.5 | Chandra, 1980 |
| Paddy-Wheat-Sesbania (GM) | Recommended for Saline-sodic soils | Mehta, 1974 |
| Paddy-Berseem-Sesbania (GM) | Well suited for alkali soils | Mehta *et al.*, 1975 |

Table 2 contd......

| Crop sequence | Desirable soil situation | Reference |
|---|---|---|
| Sesbania-Paddy-Berseem | Best suited to mild alkali soils | Kanwar and Bhumbla, 1961<br>Agarwal and Gupta, 1968 |
| Sesbania-Paddy-Barley | Well suited to moderately sodic soils | Agarwal and Gupta, 1968 |
| Sesbania-Paddy-Sugarbeet | Suitable for saline-sodic and sodic soils with pH 9 to 10 | Kanwar & Singh, 1968<br>Singh *et al.*, 1969 |
| Sesbania–Sugarcane-Cotton | Recommended for alkali soils of Deccan | Talati, 1947 |
| Sesbania-Paddy–Oats/Barley/Senji/Fenugreek | Could be adopted on highly sodic soils | Uppal *et al.*, 1962 |

nitrate, calcium chloride, potassium nitrate, potassium chloride each and 200 ppm of ascorbic acid and albumin for 24 hours before sowing. Seeds soaked in water for the same duration were used as control. The treated seeds were sown in pots having soils of different ESP (11, 25, 35 and 45) developed artificially. Their results showed that grain yield of wheat was adversely affected with increasing levels of ESP. Of all the treatments studied calcium nitrate and calcium chloride were more effective in reducing the adverse effects of sodicity in grain yield, especially towards higher ESP levels.

Treating rice seedlings with 1% solution of calcium chloride for 16 hours before transplanting resulted in increased yield by 48 and 103% at pH 9.8 and 10.1 respectively (Table 3). In a field experiment wherein 30 days old rice seedlings were dipped in various chemical solutions for 15 hours before transplanting, it was observed that grain yields increased by 1.19 to 2.02 times at pH 10.0 over that of control. Maximum increase in yield was observed in calcium chloride followed by calcium nitrate treatments (Quadar *et al.*, 1980).

Table 3 : Effect of seedling treatments on grain yield (t/ha) of rice (Var. Jaya) at different pH levels.

| Treatments | pH | | | | |
|---|---|---|---|---|---|
| | 8.5 | 9.5 | 9.8 | 10.0 | Mean |
| Control | 6.911 | 6.772 | 3.998 | 1.693 | 4.843 |
| $Ca(NO_3)_2$ 1% | 7.257 | 7.088 | 5.766 | 2.583 | 5.673 |
| $Ca(NO_3)_2$ 2% | 7.102 | 6.681 | 4.763 | 3.136 | 5.452 |
| $CaCl_2$ 1% | 7.241 | 7.291 | 5.931 | 3.436 | 5.976 |
| $CaCl_2$ 0.5% | 9.815 | 7.025 | 5.336 | 3.356 | 5.638 |
| Mean | 7.064 | 6.997 | 5.164 | 2.840 | |
| C.D. (5%) for pH | | | = 0.284 | | |
| for chemical effect | | | = 0.317 | | |
| for interaction | | | = 0.635 | | |

Source : Quadar *et al.* (1979)

Fertile tillers and number of grains per panicle were adversely affected by high ESP (Pearson and Bernstein, 1950; Joshi, 1976).

These two characters were less affected in the case of calcium nitrate and calcium chloride treated seedlings of rice. Calcium is known to maintain the integrity of the plant-cell membranes and hence to prevent the free diffusion of potentially toxic sodium ions present in a sodic environment into the cytoplasm (Chaudhuri and Wiebe, 1968, Rains, 1972). Besides, calcium and sodium have an antagonistic effect on each other and calcium also helps the plants to take up potassium (Epstein and Hagen, 1952, Ayoub, 1974). Thus calcium taken up by the seedlings during the treatments possibly help to maintain the integrity of the membrane of the root cells and also to prevent the quick accumulation of sodium in the plant system and at the same time promote the uptake of potassium,which help to maintain the chemo- osmotic potential of the cells. These two factors give initial advantage to the seedlings over the control seedlings in sodic soils, leading to better crop growth and higher grain yields.

Forgoing discussion clearly indicate that alkali tolerance in plant can also be induced to some extent by chemical treatment of seeds or seedlings before sowing or transplanting, in addition to selection of suitable crops and varieties for high alkali conditions.

## Method of Sowing

In a field experiment with maize (Ganga-5) on a saline alkali soil (EC 5.0 mmhos/cm, pH 9.8, ESP 92) with four treatments, viz. flat method of sowing; sowing on the sides of the ridge within the capillary fringe just above the irrigation water level in the furrow; sowing at the bottom of the furrow and side of the ridge with irrigation in alternate furrows; and sowing in beds 2 m apart. Results of this experiment (Rai and Mehrotra, 1971) show that sowing on the sides of the ridges provided the highest yield. In a trial on sugarbeet, Dargan and Chillar (1973) found that in highly sodic soil, sowing the seed on one side of the ridge half way between top and bottom opposite to the sun direction gave significantly higher yield than that in other methods i.e. flat sowing, alternate ridge sowing, transplanting on flat surfaces, and transplanting on ridges (Table 4). In sugarcane, Dargan *et al.* (1973) observed that sowing in trenches resulted in significantly higher yield than that in flat method of sowing in a sodic soil having pH

10.6 (Table 5). In trench sowing, ridges were made 65 cm apart and trenches were dug in furrows when sown after applying FYM and gypsum.

Table 4 : Effect of method of sowing on the yield of sugarbaet

| Treatment | % of plants survived | Yield of root (q/ha) |
|---|---|---|
| Flat sowing | 37.5 | 128.0 |
| Ridge sowing | 71.6 | 194.0 |
| Alternate ridge sowing | 67.8 | 157.8 |
| Transplanting on flat surfaces | 58.2 | 133.8 |
| Transplanting on ridges | 84.3 | 137.1 |

Source : Dargan and Chillar (1973)

Table 5 : Effect of method of sowing sugarcane on the yield (q/ha)

| Amendments (t/ha) | Yield (q/ha) | |
|---|---|---|
| Gypsum + FYM | Flat sowing | Trench sowing |
| 9 + 25 | 153.0 | 264.5 |
| 2 + 30 | 280.4 | 435.0 |
| 18 + 25 | 285.8 | 405.7 |
| 18 + 50 | 352.7 | 433.6 |

Source : Dargan et al (1973)

## Plant Population

Either too dense or too thin plant population leads to substantial loss in potential yield of field crops. An optimum plant population with proper spacing gives maximum yield in a given environment which is modified by, among other factors by soil chemical and physical conditions, fertility level and moisture regime. Alkali menance of soil which represents its chemical condition inhibit seed germination and restricts tiller production in cereals resulting in poor crop stand and reduced yield. Singh (1983) investigated the seed rate and spacing for wheat in an alkali soil (pH 10.2) incorporated with gypsum at the rate of 6 t/ha and 12 t/ha so as to create two soil types namely poorly reclaimed and well reclaimed. Treatments consisted of four seed rates (80, 100, 120 and 140 kg/ha) and 3-row spacings (that is 15, 20 and 25 cm). The row spacings served as main plot and seed rates as subplot. The results show that higher seed rate (120 kg/ha) with a narrow spacing in poorly reclaimed and a wider spacing in well reclaimed soil provided the maximum yields.

Dargan *et al.* (1974) reported that in highly sodic soil, the grain yield of rice is increased significantly with the increases in number of plants per hill (Table 6).

Table 6 : Grain yield of rice crop (q/ha) due to varieties and plant population

| Varieties | Plants per hill | | | Mean of varieties |
|---|---|---|---|---|
| | 2 | 4 | 6 | |
| IR 8–68 | 40.7 | 46.8 | 57.9 | 48.3 |
| IR 8 | 27.5 | 38.9 | 48.0 | 38.1 |
| Jhona 349 | 27.8 | 35.1 | 33.3 | 32.1 |
| Jhona 351 | 23.1 | 29.2 | 34.2 | 28.8 |
| Jaya | 15.2 | 33.0 | 36.8 | 28.3 |
| B–370 | 21.6 | 24.7 | 27.2 | 29.4 |
| B–217 | 16.0 | 20.8 | 26.6 | 21.4 |
| Mean | 24.6 | 32.5 | 37.6 | |

| CD (5%) | | |
|---|---|---|
| | Varieties | 8.4 |
| | Plant population | 3.3 |
| | Interaction | 8.8 |

Source : Dargan *et al.* (1974)

## Puddling and Pre-plant Submergence

Although rice is particulary not alkali tolerant, but because of the specific conditions of its growth and nature of the plant, it often grows well on sodic soils. In addition to the production of good yields, the standing water also helps in the amelioration of these soils. However, for getting good yields of this crop in sodic soil, the agronomic practices have to be different. Pudding of soil before transplanting of rice is a common practice. The main advantages of this practice are lower parcolation losses of water and control of weeds. Sodic soils are known for their low permeability and that the weed problem is also not serious in such soils. Under such conditions Dargan (1979) found puddling to be a wasteful practice (Table 7). However, after 4-5 years of cropping when the pH of the soil goes down around 8.5, puddling may give beneficial response.

Table 7 : Effect of puddling on yield of rice crop

| Treatments | Grain yield (t/ha) |
|---|---|
| Puddling | 12.5 |
| No Puddling | 21.5 |
| C.D. at 5% | 7.8 |

Source : Dargan (1979)

Swarup (1980) investigated the effect of submergence prior to transplanting with and without FYM on the yield and nutrition of rice in sodic soil reclamation. His results indicate that the maintenance of flooded condition for 15 days with FYM and 30 days with as well as without FYM were beneficial for rice. This beneficial effect is ascribed to the fact that calcareous sodic soils undergo reduction in pH within 1-2 weeks of flooding (Ponnamperuma, 1972) which increases the availability of most plant nutrients to the crop. Moreover, due to the reduced condition of the soil, the availability of Fe and Mn, which are required by rice plant in higher amounts (Aiyer, 1948; Sankaram, 1969) is substantially increased. On the other hand, rice plants transplanted immediately after flooding showed lower yields due to high pH and high ESP of sodic soil besides reduced availability of nutrients. The addition of organic matter accelerates the reduction condition of the soil and increase the availability of many nutrients to the crop (IRRI, 1963, 1964). The degree of improvement in sodic soil increased with the increase in the duration of submergence of calcareous sodic soils leads to the accumulation of $CO_2$ (Ponnamperuma, 1972) which solubilises the native $CaCO_3$ and this in turn may replace the adsorbed Na from the exchange complex. Submergence alongwith addition of organic matter hastens the process of reclamation possibly due to more production of $CO_2$ and organic acids.

## Irrigation with Sodic Waters

Continuous use of irrigation waters having high RSC tends to precipitate Ca and Mg in the form of carbonates. This leads to dominance of sodium carbonate and bicarbonate in soil solution and of Na on the exchange complex which replaces the Ca and accentuates the problem of sodicity in the soil. The first attempt should be to go for deep boring upto 150 metres as high RSC waters are usually met in the shallow aquifers (Mehta, 1983). If an alternative source of good quality water is available, blending water supplies can also reduce the sodicity of water. The composition of blended water could be worked out using following formula :

| | | |
|---|---|---|
| Resulting me/L | | [me/L of (a) x proportion of (a)] + |
| of Ca, Mg or | = | [me/L of (b) x proportion of (b) ] |
| RSC of blend | | |

The ground water should be treated with gypsum or sulphuric acid if the above approaches are not feasible. Gypsum in powdered form can be applied to soil before every irrigation. There is no need to apply gypsum if the RSC of irrigation water is less than 2 me/1 (Anonymous, 1972). However, addition of 850 Kg/ha of gypsum need to be added for correcting each me/1 of RSC for one hectare metre of water. Alternatively high RSC waters could be treated by placing gypsum lumps/clods/powder in permeable bags in the water course in the form of gypsum-beds (Ahmed *et al.* 1979; Pal and Poonia, 1979). Treatment of high RSC water by passing it through gypsum-bed saves the cost and labour involved in grinding gypsum and applying it on the surface of the soil.

Sulphuric acid applied to irrigation water removes carbonate and bicarbonate and reduces Ca precipitation and Na accumulation in soils (Bohn and Westerman, 1971, Miyamoto *et al.* 1975a). When bicarbonate concentration becomes sufficiently low, Ca dissolves from $CaCO_3$ as acid treated water flows into calcareous soils, even if the pH of irrigation water is above neutrality. If acid application rates exceed the bicarbonate content, the pH of irrigation water decreases drastically causing corrosion of concrete ditches and metal pipes. Field practice of irrigation water treatment is a simple one, requiring an acid tank with a dripper (Miyamoto and Stroehlein, 1975). Wen surface irrigation systems contain concrete ditches or metal pipes, acid rates are limited to about 90% of the equivalent amount of bicarbonate and carbonate in order to avoid corrosion. No standard rate of acid has yet been established for sprinkler systems. The amount of acid needed for irrigation water treatment ranges from 50 to 200 ppm or 0.5 to 2 metric tons per hectare metre of water.

## CONCLUSIONS

1. Mulching improves the physical and microbiological properties of sodic soils and helps in the leaching of soluble sodium salts.

2. Deep tillage helps in turning onto the surface B-horizon containing $CaCO_3$ and $CaSO_4$ which may bring about self amelioration.

3. Selection of tolerant crops and suitable crop rotation not only provides attractive returns from sodic soils, but also accelerates the process of reclamation.

4. To some extent, alkali tolerance could also be inducted by treating the seeds/seedlings with salt solutions before planting.

5. Green manuring with dhaincha is effective in improving the physical and chemical properties of alkali soils besides improving their nitrogen status.

6. Sowing on the sides of the ridges opposite to the direction of sun for maize and sugarbeet and trench sowing for sugarcane provides better results.

7. Higher seed rate/higher number of seedlings per hill with narrower spacing results in higher yield of crops on sodic soils.

8. Puddling of sodic soil is a wasteful practice for rice cultivation. However, pre-plant submergence for 15-30 days promotes better growth.

9. The adverse effects of sodic irrigation water could be nullified by mixing gypsum or by neutralizing its alkalinity by treating it with $H_2SO_4$. If an alternative source

of good quality water is available, blending of water supplies may also be useful.

## REFERENCES


Abrol, I.P. and Bhumbla, D.R. 1979. Crop response to differential gypsum applications in a highly sodic soil and the tolerance of several crops to exchangeable sodium under field conditions. *Soil Sci.* 127 : 79-85.

Abrol, I.P. and Fireman, D.R. 1977. Alkali and Saline Soils. *Tech. Bull,* CSSRI, Karnal, No.4 pp. 32.

Acharya, C.L. and Abrol, I.P. 1976. Effect of river sand on the permeability of a sodic soil. *J. Indian Soc. Soil Sci.,* 24 : 245-252.

Agarwal, R.R., Yadav, J.S.P. and Gupta, R.N. 1979. Saline and Alkali Soils of India. ICAR, New Delhi, p. 286.

Ahmed, B., Kemper, W.D., Ghulam, H. and Niazi, M.A. 1979. Use of gypsum stones to lower the sodium adsorption ratio of irrigation water. *Soil Sci. Soc. Am. J.* 43 : 698-702.

Aiyer, S.P. 1948. The place of minor elements in rice growing. *Indian Fmg.* 7 : 11-14.

Anjaneyulu, B.S.R. 1972. Drainage improvement for reclamation and preventing water logging in some soils of Rajasthan canal area. *Cent. Bd. Irrig. Pwr. Pub.* 118(2) : 43-48.

Anonymous 1972. Proceeding of the meeting on water quality rating in India. CSSRI, Karnal.

Antipov-karatov, I.N., Yurin, I.A. and Frolkina, L.A. 1957. Reclamation of solonetz in the chernozem zone-results of experiments in Kamiennaya steppe. *Pochvovedenie* 2 : 1-17.

Asghar, A.G. and Khan, H. 1958. Field studies in prevention of soil salinization. *Agron. J.* 50 : 667-671.

Ayoub, A.T. 1974. Effect of calcium on sodium salinization of beans (*Phaseolus vulgaris*). *J. Expt. Bot.,* 25 : 245-252.

Bains, S.S. 1971. Agronomic investigations in utilization of waste lands in saline and alkali soils. *Proc. Symp. Sci. & India's Fd. Probl. New Delhi, 1967,* p. 198-203.

Bains, S.S., Daya Nand and Singh, K.N. 1969. Keep soil salinity at bay with continuous cropping. *Indian Fmg.,* 19(5) : 7-8.

Bajpai, P.D., Gupta, B.R. and Singh, C. 1976. Some microbiological studies in salt-affected soils. II-Effect of gypsum and mulch on selected soil microflora and evolution of carbon dioxide. *J. Indian Soc. Soil Sci.* 24 : 48-52.

Balyabo, N.K. 1957. The ameliorative layer-ploughing as principal method of reclaiming steppe solonetz and solonetz-like soils. *Pochvovedenie* 2 : 18-30.

Balyabo, N.K.; Gutina, B.S. and Blincv, M.I. 1956. The importance of 3-layer plowing in reclamation of steppe-solonetz and solonetz-like soils. *Byull, Vses. nauchno-issled. Udobr. Agrotekh-Agropochy,* 1 : 28-31.

Bandopadhya, A.K. and Sen, H.S. 1977. Introduction of a second rice crop and mulching as a reclamatory management practices of saline soils of mono-cropped areas of South 24-Parganas, West Bengal, *J. Indian Soc. Soil Sci.* 25 : 326-329.

Basu, J.K. and Tagare, V.D. 1943. A note on the alkali soils of the Bombay State, India. *Trans. 4th int. Congr. Soil Sci.* 3 : 169-172.

Bhardwaj, K.K.R., Venkat Krishnan, S. and Batra, L. 1977. Studies on manurial value of *dhaincha* as a green manure crop. *Ann. Rep., CSSIR, Karnal,* p. 36.

Bhumbla, D.R. 1972. Reclamation of saline and sodic soils. *Indian Fmg.* 32(2) : 19-21.

Bohn, H.L. and Westerman, R.L. 1971. Sulfuric acid : Its potential for improving irrigation water quality, *Proc. Am. Water Res. Assoc., Ariz. Sect.,* 1 : 42-52.

Botov, T.G. 1959. Improvement of solonetzic and podzolic soils by redistribution of the soil horizons. *Pochvovedenie,* 5 : 80-86.

Bull, M.R. 1962. Combined beet factory-cum-refinery : first sugar factory in Tunisia. *Sugar Arucar, Puerto Rico,* p. 58.

Cairns, R.R. 1969. Canadian solonetz soils and their reclamation. *Agrokem. talajt.* 18(Supp1) : 233-237.

Chahal, R.S. 1962. Removal of salts from saline-alkali soils by use of porous ducts, *J. Indian Soc. Soil Sci.* 19 : 289- 293.

Chandra, S. 1980. Evaluations and breeding of barley varieties suited to salt affected soils. *Ann. Rep., CSSRI, Karnal.*

Chaudhuri, I.I. and Wiebe, H.H. 1968. Influence of calcium pretreatment on wheat germination in saline media. *Pl. Soil,* 28 : 208-216.

Dargan, K.S. 1979. Agronomic and cultural practices for crop production in salt affected soils. *Curr. Agric.*, 3 : 1-20.

Dargan, K.S. and Chillar, R.K. 1973. A technique for sugarbeet and vegetable in saline-alkali soils. *Indian Fmg.*, 22(12) : 13-16.

Dargan, K.S. and Chillar, R.K. 1980. Studies on green-manuring and nitrogen requirements of rice and their residual effect on wheat on a sodic soil. *Int. Symp. Salt Affected Soils, Karnal* pp. 440-446.

Dargan, K.S., Chillar, R.K. and Mehta, K.K. 1974. In saline- alkali soils, demonstrations prove effectiveness of new practices. *Indian Fmg.*, 23(10) : 11-13.

Dargan, K.S., Gaul, B.L. and Abrol, I.P. 1973. Growing sugarcane in saline-alkali soils. *Indian Fmg.*, 23(2) : 8-10.

Dhawan, C.L., Bhatnagar, B.B.L. and Ghai, P.D. 1958a. Role of green-manuring in reclamation. *Proc. natn. Acad. Sci. India*, 27A :168-176.

Dhawan, C.L., Bhatnagar, S.B.L. and Singh, J. 1958 b. Reclamation of alkali soils by different crop rotations. *Proc. nat. Acad. Sci. India* 27A : 238-252.

Dudanov. M.F. 1954. A note on the improvement of saline soils. *Nauch. Tr. Lab. Poch. Akad. Nauk. Ukrain SSR* 1 : 84-94.

Dwivedi, R.S. 1979. Plant Physiology, *In* A decade of Research. CSSRI, Karnal, pp. 99-116.

Epstein, E. and Hagen, C.E. 1952. A kinetic study of the absorption of alkali cations by barley root. *Plant Physiol.*, 27 : 457-474.

I.R.R.I. 1963. *Annual Report*. The International Rice Research Institute, Philippines.

I.R.R.I. 1964. *Annual Report*. The International Rice Research Institute, Philippines.

Johnson. W,E. 1979. Land reclamation in Saskatchewan agriculture. *Proc. 4th Ann. Meeting, Canadian Land Reclamation Assoc. Regins*, p. 1-13.

Joshi, Y.C. 1976. Effect of different levels of ESP on the yield attributes of seven wheat veriaties. *Indian J. Plant Physiol.*, 19 : 190-193.

Joshi, Y.C., Qadar, A. and Sharma, S.K. 1981. Alkali halophytes tolerance of *Desmostachya bipinnata. Curr. Agric.*

Kanwar, J.S. 1962. Green-manuring and reclamation of saline- alkali soils. *Proc. Sem. Salinity Alkali Soil Probl., New Delhi*, pp. 68-77.

Kanwar, J.S. 1969. Sugarbeet for saline and alkali soils of Northern India. *Indian Fmg.* 19(2) : 5-6.

Kanwar, J.S. and Bhumbla, D.R. 1961. Annual Progress Report of the Scheme on reclamation of Saline-alkali soils in the Punjab. PAU, Ludhiana.

Kanwar, J.S. and Singh, N.T. 1968. Sutability of sugarbeet for saline alkali soils of Northern India. I- Adaptability. *Indian J. agric. Sci.*, 38 : 115-121.

Khosla, B.K. and Yadav, J.S.P. 1976. Rice husk improves physical condition and crop growth in sodic soils. *Indian Fmg.*, 25(10) : 9-11.

Kumar, A. and Abrol, I.P. 1979. Performance of five perennial forage grasses as influenced by gypsum levels in a highly sodic soil. *Indian J. agric. Sci.*, 49 : 473-477.

Kumar, A., Abrol, I.P. and Dargan, K.S. 1980. Karnal grass grows well in sodic soils. *Indian Fmg.*, 30(3) : 14-15.

Kumar, A., Abrol, I.P. and Dargan, K.S. 1981. Shaftal, a promising fodder for alkali soils. *Indian Fmg.* 31(4) : 18- 20.

Maksimyuk, G.P. 1958. The salt regime of the solonchak solonetz soils and its change during reclamation. *Akad. Nauk, SSSR Inst. Lesa. Trudy* 33 : 83-93.

Mehta, K.K. 1974. Grow bumper crops in alkali soils. *Inf. Bull.* No. 1, CSSRI, Karnal.

Mehta, K.K. 1983. *Reclamation of Alkali Soils in India.* Oxford & IBH Publishing Co., New Delhi, pp. 280.

Mehta, K.K.; Yadav J.S.P. and Abrol, I.P. 1975. In village Kachwa, alkali fields turn productive. *Indian Fmg.*, 25(9) : 17-20.

Milap Chand, Abrol, I.P. and Bhumbla, D.R. 1977. Comparision of the effect of eight amendments on the soil properties and crop growth in a highly sodic soil. *Indian J. agric. Sci.*, 47 : 348-354.

Mishra, B.; Gill, K.S.; Sarin, M.N. and Singh, T.N. 1973. Sugarbeet will do well in saline-alkali lands. *Indian Fmg.*, 23(4) : 29-30.

Miyamoto, S. and Stroehlein, J.L. 1975. Sulfuric acid for increasing water penetration into some Arizona soils. *Prog. Agric. Ariz.*, 27 : 13-16.

Miyamoto, S., Ryan, J. and Stroehlein, J.L. 1975. Potentially beneficial uses of sulfuric acid in Southwestern agriculture. *J. Environ. Qual.* 4 : 431-437.

Nadezhdin, B.V. 1957. Characteristics of salinized soils of the middle Trans-Ural region. *Pochvovedenie* 6 : 86-94.

Novikova, A.V. 1956. Ergebnisse and perspektiven der Melioration von Alkaliboden in der Krim. *Pochvovedenie* 8 : 31-43.

Novikova, A.V. 1961. Specific features of the amelioration of meadow solonetzes. *Nauch. Tr. Ukrain. nauch-issled Inst. Pochvoved* 5 : 121-131.

Pak, K.P. 1957. Methods of increasing the fertility of solonetz soils. *Timiriazevskaia Sel'skokhoz. Akad. Izv.* 5 : 63-90.

Pal, R. and Poonia, S.R. 1979. Dimensions of gypsum bed in relation to residual sodium carbonate of irrigation water, size of gypsum fragments and flow velocity. *J. Indian Soc. Soil Sci.,* 27 : 5-10.

Pearson, G.A. and Bernstein, L. 1958. Influence of exchangeable sodium on yield and chemical composition of plants. II-Wheat, barley, oats, rice, tall fesue and tall wheat grass. *Soil Sci.,* 86 : 254-261.

Petrosian, G.P. 1956. The role of lucerne in the reclamation of salinized soils. *Izv. Akad. Nauk. Armyan SSR biol. S-kh Nauki,* 12 : 77-89.

Ponnamperuma, F.N. 1972. The chemistry of submerged soils. *Adv. Agron.* : 29-96.

Prettenhoffer, I. 1963. Experimental results on reclamation of alkali (szik) grassland in the region beyond the Tisza river. VI- Demonstrative trials on grassland amelioration in Csongrad country (1958-1961). *Kiserl. Kozl.* 56A : 19-44.

Prettenhoffer, I. 1964. Advances in the amelioration of carbonate free alkali soils (meadow solonetz) by sub-soiling. *Agrokem. Talajt.* 13 (Suppl.) : 227-235.

Prettenhoffer, I. 1974. Amelioration and utilization of solonetzes in the region east of the river Tisza. *Agrokem. Talajt.* 23(Suppl.) : 181-190.

Quadar, A., Joshi, Y.C., Bal, A.R. and Dwivedi, R.S. 1979. Induction of salt resistance in crop plants by chemical treatments. *Ann. Rep. CSSRI, Karnal,* pp. 95-96.

Quadar, A., Joshi, Y.C., Bal, A.R. and Dwivedi, R.S. 1980. Effect of calcium salts on induction of sodicity resistance in rice. *Curr. Agric.* 4 : 83-86.

Quadar, A., Joshi, Y.C., Bal, A.R. and Rana, R.S. 1980. Chemical treatments to enhance cereal production in salt affected soils. *Intern. Symp. Salt Affected Soils, Karnal,* 468-471.

Rabochev, I. 1954. The improvement of saline soils in the Amu- Darya zone. *Khlopkovodstvo,* 12 : 47-50.

Rai, B. and Mehrotra, C.L. 1971. Agronomic methods for minimizing the salt hazards on the germination and growth of crops. *All India Symp. Soil Salinity, Kanpur,* pp. 212-214.

Rains, D.W. 1972. Salt transport in relation to salinity. *Ann. Rev. Plant Physiol.,* 33 : 367-388.

Rana, R.S. and Prakash, V. 1980. Karnal grass grows well in alkali soils. *Indian Fmg.,* 30(4) : 17-19.

Rana, R.S., Singh, K.N., Prakash, V. and Ahuja, P.S. 1980. Growing forage in barren alkali soils. *Seeds Farms,* 6(1) : 37-39.

Sambur, G.N. 1954. The question of the fundamental improvement of solonetzes of the dry steppe of the USSR. *Pochovedenie* 7 : 103-109.

Sandoval, F.M. 1978. Deep plowing improves sodic claypan soils. *N. Dak. Fm. Res.,* 35 : 15-18.

Sankarram, A. 1969. Biochemical transformation and nutrition of rice under waterlogged conditions- A review. *Indian J. agric. Sci.,* 39 : 299-323.

Satyanarayan, M. 1953. Crop and land improvement under the Krishna Kast Bank Canal (Madras State). *Madras agric. J.* 40 : 293-294.

Semenova-Zabrodina, S.P. 1952. Cultivation of solonetzes of Proshivash by overturning. *Pochvovedenie,* p. 426-433.

Sharma, D.P., Mehta, K.K.; Yadav, J.S.P. and Gurucharan Singh (1983). Effect of crop rotation on crop growth and soil properties in sodic soils on farmer's fields. *Indian J. agric. Sci.,* 53 : 681-685.

Singh, N.D., Kanwar, J.S. and Bhumbla, D.R. 1969. Suitability of sugarbeet for saline and saline-alkali soils of northern Indian. II-Economic possibilities. *Indian J. agric. Sci.,* 39 : 122-128.

Singh, T.N. 1983. Increasing wheat yield from alkali soils in eastern Uttar Pradesh. *Indian Fmg.* 33(8) : 27-28.

Sipos, S. 1975. Interaction between physical properties and fertility of solonetz soils. *Int. Symp. New Develop. Field Salt Affected Soils, Cairo, 1972,* p. 691-696.

Sokolovskii, S.P. 1956. Soil-reclamation regionalization of the territory of the Nogai Steppe. *Pochvovedenie* 11 : 68-72.

Somani, L.L. 1981. Reclamation of a saline-alkali soil using organic materials and inorganic amendments. *Agrokem. Talajt.* 30 : 333-350.

Somani, L.L. and Saxena, S.N. 1981. Role of some sources of organic materials alone and in conjuction with inorganic amendments in reclamation of a typical calcareous saline-alkali soil of Rajasthan. I-Effects on soil physical properties and crop yield. *Agrochimica* 25 : 232-242.

Somani, L.L. and Saxena, S.N. 1982. Role of some sources of organic materials alone and in conjuction with inorganic amendments in reclamation of a typical calcareous saline alkali soil of Rajasthan. II-Effect on soil chemical properties and crop yield. *Agrochimica* 26 : 55-63.

Swarup, A. 1980. Effect of submergence and farmyard manure application on the yield and nutrition of rice and sodic soil reclamation. *J. Indian Soc. Soil Sci.* 28 : 532-534.

Talati, R.P. 1947. Field experiments on reclamation of salt lands in Baramati of Bombay Deccan. *Indian J. agric. Sci.* 17 : 153-174.

Tiwari, K.N.; Tiwari, S.P. and Pathak A.N. 1980. Studies on green manuring of rice in double cropping system in a partially reclaimed saline-sodic soil. *Indian J. Agron.*, 25 : 136- 145.

Totawat, K.L. and Singh, R.M. 1978. Efficacy of soil conditioners and nitrogen under saline-sodic irrigation on sugarbeet (*Beta vulgaris* L.) dry matter, yield and sugar production. *Ann. Arid Zone*, 17 : 209-215.

Uppal, H.L. 1955. Green-manuring with special reference to *Sesbania aculeata* for treatment of alkaline soils. *Indian J. agric. Sci.*, 25 : 211-235.

Uppal, H.L.; Agarwal, R.R. and Kibe, M.M. 1961. Reclamation of Saline and Alkaline Lands. Farm Bull No. 66 : 1-30. Farm Information Unit, Directorate of Extension, Ministry of Food and Agriculture, New Delhi.

Venkat Krishnan, S. 1980. Mineralization of green manure (*Sesbania aculeata*) nitrogen in sodic and dreclaimed soils under flooded conditions. *Pl. Soil*, 54 : 149-152.

Yadav, J.S.P. and Agarwal, R.R. 1959. Dynamics of soil changes in the reclamation of saline-alkali soils of the Indo-Gangatic alluvium. *J. Indian Soc. Soil Sci.* 7 : 213-222.

Yadav, J.S.P. and Agarwal, R.R. 1961. A comparative study on the effectiveness of gypsum and *dhaincha (Sesbania aculeata)* in the reclamation of a saline-alkali soil. *J. Indian Soc. Soil Sci.,* 9 : 151-156.

Zvereva, K.A. 1958. Different methods of reclamation for different solonetzes. *Doke. Akad. S-kh. Nauk,* 10 : 35-41.

Dryland Resources & Technology (1992)
Editors : L.L. SOMANI, R.L. SHYAMPURA
& M.C. MUNDRA


# Character Association, Path Analysis and Selection Indices in Foxtail Millet (*Setaria italica* (L.) Beauv)

*S.R. Maloo, R.P. Gupta and U.C. Sharma*
Department of Plant Breeding and Genetics,
Rajasthan Agricultural University,
Udaipur – 313 001 (India)

## ABSTRACT

In the present experiment an attempt was made to determine correlation, path coefficient analysis and selection indices in *Setaria italica* (L.) Beauv for grain yield, its component traits, grain protein and oil contents. Association studies revealed that, in general, there was a close agreement between genotypic and phenotypic correlation coefficients with the higher magnitude of genotypic correlations. Yield/plant was strongly correlated with plant height, panicle length, weight of panicle, yield/panicle, 1000-seed weight, biological yield and harvest index at both genotypic and phenotypic levels. However, quality traits viz., grain protein and oil content exhibited negative and significant correlation with yield/plant at both levels.

Path analysis study indicated maximum direct effects of harvest index followed by biological yield, weight of panicle and grain protein content. The positive indirect effects of most of the characters were again largely through biological yield followed by harvest index and weight of panicle.

The discriminant function and selection indices studies revealed highest relative efficiency over straight selection for yield/plant by an index which included yield/plant, harvest index, biological yield and weight of panicle. Since these parameters are easy to measure, they can serve as selection criteria in isolating superior genotypes even in early segregating generations of this important small millet crop.

## INTRODUCTION

Under the broad classification of coarse cereals finger millet (ragi) and other small millets which include foxtail millet, little millet, proso millet, kodo millet and barnyard millet. Among these, foxtail millet (*Setaria italica* (L.) Beauv) locally called kangni (kakun) is perhaps the oldest minor millet crop. This is a short duration crop of the arid regions. It is cultivated for grain as well as fodder in many parts of the world viz., China, Eastern Europe, Southern parts of Russia, India and to some extent in African and American countries.

In India, foxtail millet is cultivated under diverse situations of soil, temperature, rainfall and occupies agricultural nitches where options for crop diversification are few. Often kangni is confined to hill slopes, undulating fields and gravelly soils but still provides assured harvest of reasonable level under marginal cultivation to the poor farmers of Karnataka, Andhra Pradesh, Madhaya Pradesh, Uttar Pradesh, Bihar and Southern Rajasthan. This crop has been the source of food supply for a very large section of the rural population during famines, dry years and emergency simultaneously providing fodder for cattle.

Although foxtail millet plays a significant role in dry land farming providing food and fodder yet the yield levels of this crop are very low as compared to cereals. The probable constraints that limit its production are many and varied. The major ones appear to be non-availability of improved varieties, poor cultivation practices and lack of organised and concentrated research efforts.

The genetic improvement of a crop can be achieved by breeding high yielding varieties with improved quality. Association studies for grain yield, its components and various quality characters become very important when the breeder is confronted with the problems of combining high yield potential with desirable agronomic and quality characters. Correlation studies thus would provide reliable information on nature, extent and direction for selection. Path coefficient analysis further permits the partitioning of correlation coefficients into direct and indirect causes. In addition to this, discriminant function provides suitable selection indices to maximize the genetic advance in breeding work. Very limited research work is reported in foxtail millet on the aforesaid biometrical parameters (Sandhu *et al.* 1974; Chinnaswamaiah 1975; Dhagot 1977; Reddy and Lakshmi 1991). Therefore, the present study was undertaken to furnish the requisite information on nature of relationships among quantitative and quality parameters and their association with grain yield and relative contribution of each of these parameters to the final built-up of grain yield in foxtail millet.

## MATERIALS AND METHODS

The experimental material for the present study comprised 40 diverse varieties/strains of *Setaria italica* (L.) Beauv raised in randomised block design with three replications during *kharif,* 1991. Each entry was sown in two lines per replication in a row length of 3m. Row to row and plant to plant distances were maintained as 25 x 7.5 cm respectively. Fertilizers were applied at the rate of 40 N : 20 $P_2O_5$ kg/ha. Half dose of nitrogen was applied at the time of sowing while remaining was top-dressed after 30 days of sowing. Appropriate agronomical practices were adopted for raising the crop in rainfed situation.

Observations were recorded on five randomly selected plants of each variety in each replication on 15 characters viz., days to flower, days to maturity, plant height, effective tillers/plant, panicle length, peduncle length, weight of panicle, flag leaf area, yield/panicle, 1000-seed weight, biological yield/plant, yield/plant, harvest index, grain protein content and grain oil content. Genotypic and phenotypic correlations (Johnson *et al.* 1955) and path coefficient analysis (Wright, 1921 and Dewey and Lu, 1959) were worked out. Selection indices were constructed following Robinson *et al.* (1951).

## RESULTS AND DISCUSSION

### Correlation Coefficients

Genotypic and phenotypic correlations of different characters with grain yield and among themselves (Table 1) clearly indicated that, in general, there was a close agreement between genotypic and phenotypic correlation coefficients thus indicating that the environmental influences were insignificant. It was further noticed that, in general, almost all characters exhibited slightly higher genotypic correlations in comparison to their corresponding phenotypic correlations for most of the characters.

As obvious from the Table 1, yield/plant was strongly and positively correlated with plant height, panicle length, weight of panicle, yield/panicle, 1000-seed weight, biological yield and harvest index at genotypic and phenotypic levels.

Similar associations for one or other aforesaid component characters with grain yield in foxtail millet were also reported by Chinnaswamaiah (1975) for grain weight of main panicle and straw weight; Godawat and Gupta (1981) for 1000-grain weight and ear length; Wang and Han (1981) for panicle weight/plant and straw weight and Singh and Rao (1989) for straw weight, panicle weight and panicle length. While negative and significant association was observed between yield/plant and panicle length as suggested by Chinnaswamaiah (1975) in foxtail millet.

In other small millets, similar association of one or other component traits with yield was recorded by Patnaik (1968), Agolodia *et al.* (1979), Dhanakodi *et al.* (1989) in ragi; by Yadav and Srivastava (1976) in little millet and by Natarajan *et al.* (1978) and Saxena *et al.* (1979) in proso millet.

Table 1 further reveals negative and significant association of grain protein content and grain oil content with yield at genotypic and phenotypic levels. Liang *et al.* (1969) and Gasim *et al.* (1975) also reported negative assoication of protein content with grain yield in sorghum.

Variable results were obtained with respect to mutual correla-

Table 1 : Genotypic ($r_g$) and phenotypic ($r_p$) correlation coefficients between fifteen characters in foxtail millet

| Characters | | Days to flower | Days to maturity | Plant height | Effective tillers/plant | Panicle length | Peduncle length | Weight of panicle | Flag leaf area | yield/panicle |
|---|---|---|---|---|---|---|---|---|---|---|
| Days to flower | $r_g$ | – | 0.93** | – 0.04 | –0.52** | 0.10 | –0.28 | 0.14 | 0.28 | 0.13 |
| | $r_p$ | – | 0.83** | –0.03 | –0.36** | 0.09 | –0.26 | 0.13 | 0.27 | 0.10 |
| Days to maturity | $r_g$ | | – | 0.04 | –0.26 | 0.10 | –0.18 | 0.14 | 0.24 | 0.15 |
| | $r_p$ | | – | 0.04 | –0.16 | 0.09 | 0.18 | 0.13 | 0.23 | 0.13 |
| Plant height | $r_g$ | | | – | –0.14 | 0.45** | 0.44** | 0.52** | 0.02 | 0.55** |
| | $r_p$ | | | – | –0.09 | 0.45** | 0.43** | 0.47** | 0.02 | 0.49** |
| Effective tillers/plant | $r_g$ | | | | – | –0.17 | 0.11 | –0.42** | –0.53** | –0.34* |
| | $r_p$ | | | | – | –0.08 | 0.08 | –0.22 | –0.33* | –0.16 |
| Panicle length | $r_g$ | | | | | – | 0.10 | 0.57** | 0.21 | 0.57** |
| | $r_p$ | | | | | – | 0.09 | 0.54** | 0.21 | 0.52** |
| Peduncle length | $r_g$ | | | | | | – | 0.28 | 0.02 | 0.29 |
| | $r_p$ | | | | | | – | 0.24 | 0.02 | 0.26 |
| Weight of panicle | $r_g$ | | | | | | | – | 0.40* | 0.99** |
| | $r_p$ | | | | | | | – | 0.35* | 0.93** |
| Flag leaf area | $r_g$ | | | | | | | | – | 0.32* |
| | $r_p$ | | | | | | | | – | 0.29 |
| Yield/panicle | $r_g$ | | | | | | | | | – |
| | $r_p$ | | | | | | | | | – |
| 1000-seed weight | $r_g$ | | | | | | | | | |
| | $r_p$ | | | | | | | | | |
| Biological yield | $r_g$ | | | | | | | | | |
| | $r_p$ | | | | | | | | | |
| Yield/plant | $r_g$ | | | | | | | | | |
| | $r_p$ | | | | | | | | | |
| Harvest index | $r_g$ | | | | | | | | | |
| | $r_p$ | | | | | | | | | |
| Grain protein content | $r_g$ | | | | | | | | | |
| | $r_p$ | | | | | | | | | |
| Grain oil content | $r_g$ | | | | | | | | | |
| | $r_p$ | | | | | | | | | |

Table 1 Contd.............

| Characters | | 1000-seed weight | Biological yield | yield / plant | Harvest index | Grain protein content | Grain oil content |
|---|---|---|---|---|---|---|---|
| Days to flower | $r_g$ | 0.22 | 0.00 | –0.08 | –0.12 | 0.22 | –0.03 |
| | $r_p$ | 0.17 | –0.07 | –0.07 | –0.04 | 0.21 | –0.02 |
| Days to maturity | $r_g$ | 0.33* | 0.05 | 0.02 | –0.07 | 0.25 | –0.11 |
| | $r_p$ | 0.31 | 0.06 | 0.02 | –0.07 | 0.23 | –0.11 |
| Plant height | $r_g$ | 0.43** | 0.56** | 0.58** | 0.27 | –0.57** | –0.60** |
| | $r_p$ | 0.40** | 0.46** | 0.49** | 0.21 | –0.55** | –0.57** |
| Effective tillers/plant | $r_g$ | –0.10 | –0.07 | –0.15 | –0.13 | 0.08 | 0.08 |
| | $r_p$ | –0.07 | 0.08 | 0.08 | 0.04 | 0.07 | 0.09 |
| Panicle length | $r_g$ | 0.29 | 0.44** | 0.64** | 0.42** | –0.55** | –0.53** |
| | $r_p$ | 0.28 | 0.38** | 0.56** | 0.33* | –0.52** | –0.50** |
| Peduncle length | $r_g$ | 0.13 | 0.24 | 0.35* | 0.27 | –0.46** | –0.37* |
| | $r_p$ | 0.12 | 0.19 | 0.30 | 0.22 | –0.45** | –0.35* |
| Weight of panicle | $r_g$ | 0.42** | 0.67** | 0.92** | 0.57** | –0.60** | –0.89** |
| | $r_p$ | 0.35** | 0.59** | 0.83 | 0.47** | –0.53** | –0.77** |
| Flag leaf area | $r_g$ | 0.07 | 0.10 | 0.21 | 0.19 | 0.04 | –0.23 |
| | $r_p$ | 0.06 | 0.07 | 0.17 | 0.15 | 0.04 | –0.22 |
| Yield/panicle | $r_g$ | 0.48** | 0.66** | 0.96** | 0.62** | –0.63** | –0.94** |
| | $r_p$ | 0.39* | 0.60** | 0.91** | 0.55** | –0.53** | –0.79** |
| 1000-seed weight | $r_g$ | – | 0.52** | 0.44** | 0.04 | –0.16 | –0.32* |
| | $r_p$ | – | 0.43** | 0.36* | 0.02 | –0.16 | –0.30 |
| Biological yield | $r_g$ | | – | 0.70** | –0.10 | –0.29 | –0.62** |
| | $r_p$ | | | 0.62** | –0.18 | –0.24 | –0.50** |
| Yield/plant | $r_g$ | | | – | 0.64** | –0.72** | –0.97** |
| | $r_p$ | | | – | 0.64** | –0.59** | –0.76** |
| Harvest index | $r_g$ | | | | – | –0.75** | –0.69** |
| | $r_p$ | | | | – | –0.55** | –0.47** |
| Grain protein content | $r_g$ | | | | | – | 0.65** |
| | $r_p$ | | | | | – | 0.60** |
| Grain oil content | $r_g$ | | | | | | – |
| | $r_p$ | | | | | | – |

* Significant at 5% level ** Significant at 1% level

tion between different characters. Biological yield and harvest index were strongly correlated with each other. Not only this, both of these characters were also positively and significantly associated with weight of panicle, panicle length and yield/panicle at both levels. Besides this, biological yield was also strongly associated with plant height and 1000-seed wieght at genotypic and phenotypic levels. Dhagot *et al.* (1977) indicated that straw yield was positively correlated with plant height. Contrary to the present study, Xu and Zhanes (1981) reported negative and significant association between straw weight and 1000-grain weight in foxtail millet.

Similarly, yield/panicle had significant association with plant height, 1000-seed weight, weight of panicle and panicle length at both levels. Plant height showed strong correlation with panicle length, weight of panicle, peduncle length and 1000-seed weight at genotypic and phenotypic levels. Gill and Randhawa (1975) also observed strong correlation of plant height with panicle length in *Setaria*.

Positive and significant correlation coefficient was observed between weight of panicle and panicle length, flag leaf area and 1000-seed weight at both genotypic and phenotypic levels. While Xu and Zhanes (1981) reported that spike weight/plant was highly and positively correlated with days to maturity and straw weight in *Setaria*. Further, days to flower and maturity were also strongly correlated with each other, while these characters showed either negative or insignificant association with all other characters at least at genotypic levels. Gill and Randhawa (1975) also reported significant and positive association between days to flower and days to maturity in foxtail millet.

However, it is interesting to record that effective tillers/plant were either negatively correlated or showed insignificant association with other yield attributes. Both quality traits viz., grain protein and oil content exhibited strong positive correlation with each other while, negative with plant height, panicle length, peduncle length, weight of panicle, yield/panicle and harvest index at genotypic and phenotypic levels.

Therefore, from the present correlation studies, it could be concluded that grain yield showed strong correlation with plant height, panicle length, weight of panicle, yield/panicle, 1000- seed weight, biological yield and harvest index. Hence, simultaneous selection for all these traits would result in the genetic improvement of yield levels.

## Path Coefficient Analysis

Information obtained from correlation studies does not provide a clear picture of ocntribution of each component character. At the same time, as more variables are included in association studies the direct association becomes complex and important. Under such situation path coefficient analysis is useful in partitioning the direct and indirect causes of correlation and allows a detailed examination of specific forces acting to produce a given correlation. Path analysis at the same time also measures the relative importance of each causal factors. Hence, this study provides a realistic basis for allocation of weightage to each attributes in deciding a suitable criterion for genetic improvement.

In the present case path coefficients were analysed for fourteen characters using only genotypic correlation. As observed from correlation studies, plant height, panicle length, weight of panicle, yield/panicle, 1000-seed weight, biological yield and harvest index were strongly associated with yield at both genotypic and phenotypic levels. Table 2 and Figure 1 reveal direct and indirect contribution of component traits. It was obvious from the table that maximum direct effects on the yield was recorded for harvest index followed by biological yield, weight of panicle and grain protein content. Nagarajan and Prasad (1980) also reported maximum direct contribution of straw yield on grain yield in foxtail millet.

However, direct contribution of other component traits in *Setaria* were reported by Randhawa *et al.* (1978) for tiller number and ear width and Nagarajan and Prasad (1980) for plant height and productive tillers.

Other characters of the present study like days to maturity, panicle length, peduncle length and 1000-seed weight showed positive direct effects but with low magnitude. Large direct contribution of these characters were also reported by Dhagot *et al.*

Table 2 : Path analysis showing direct and indirect effects of fourteen characters on grain yield in foxtail millet

| S. No. | Characters | Direct effect | Days to flower | Days to maturity | Plant height | Effective tillers/plant | Panicle length | Peduncle length | Weight of panicle |
|---|---|---|---|---|---|---|---|---|---|
| 1 | Days to flower | –0.05 | – | 0.07 | 0.00 | 0.06 | 0.00 | 0.00 | 0.04 |
| 2 | Days to maturity | 0.07 | –0.05 | – | 0.00 | 0.03 | 0.00 | 0.00 | 0.04 |
| 3 | Plant height | –0.07 | 0.00 | 0.00 | – | 0.02 | 0.00 | 0.00 | 0.14 |
| 4 | Effective tillers/plant | –0.12 | 0.03 | –0.02 | 0.01 | – | 0.00 | 0.00 | –0.11 |
| 5 | Panicle length | 0.01 | –0.01 | 0.01 | –0.03 | 0.02 | – | 0.00 | 0.15 |
| 6 | Peduncle length | 0.01 | 0.01 | –0.01 | –0.03 | –0.01 | 0.00 | 0.00 | 0.07 |
| 7 | Weight of panicle | 0.27 | –0.01 | 0.01 | –0.03 | 0.05 | 0.00 | 0.00 | – |
| 8 | Flag leaf area | –0.10 | –0.01 | 0.02 | 0.00 | 0.06 | 0.00 | 0.00 | 0.11 |
| 9 | Yield/panicle | –0.63 | –0.01 | 0.01 | –0.04 | 0.04 | 0.00 | 0.00 | 0.27 |
| 10 | 1000–seed weight | 0.05 | –0.01 | 0.02 | –0.03 | 0.01 | 0.00 | 0.00 | 0.11 |
| 11 | Biological yield | 1.09 | 0.00 | 0.00 | –0.04 | 0.01 | 0.00 | 0.00 | 0.18 |
| 12 | Harvest index | 1.14 | 0.01 | –0.01 | –0.02 | 0.01 | 0.00 | 0.00 | 0.15 |
| 13 | Grain protein content | 0.23 | –0.01 | 0.01 | 0.04 | –0.01 | –0.01 | 0.01 | –0.16 |
| 14 | Grain oil content | –0.04 | 0.00 | –0.01 | 0.04 | –0.01 | 0.00 | 0.00 | –0.24 |

Table 2 Contd.........

| S.No. | Flag leaf area | Yield/panicle | 1000–seed weight | Biological weight | Harvest index | Grain protein content | Grain oil content | Genotypic correlation with grain yield/plant |
|---|---|---|---|---|---|---|---|---|
| 1 | –0.03 | 0.00 | 0.01 | 0.00 | –0.14 | 0.05 | 0.00 | –0.08 |
| 2 | –0.02 | –0.09 | 0.02 | 0.05 | –0.08 | 0.06 | 0.00 | 0.02 |
| 3 | 0.00 | –0.35 | 0.02 | 0.61 | 0.30 | –0.13 | 0.02 | 0.58** |
| 4 | 0.05 | 0.22 | –0.01 | –0.07 | –0.15 | 0.02 | 0.00 | –0.15 |
| 5 | –0.02 | –0.36 | 0.02 | 0.48 | 0.48 | –0.13 | 0.02 | 0.64** |
| 6 | 0.00 | –0.18 | 0.01 | 0.26 | 0.30 | –0.11 | 0.01 | 0.35** |
| 7 | –0.04 | –0.62 | 0.02 | 0.73 | 0.65 | –0.14 | 0.03 | 0.92** |
| 8 | – | –0.20 | 0.00 | 0.11 | 0.22 | 0.01 | 0.01 | 0.21 |
| 9 | –0.03 | – | 0.03 | 0.73 | 0.70 | –0.15 | 0.03 | 0.96** |
| 10 | –0.01 | –0.31 | – | 0.57 | 0.04 | –0.04 | 0.01 | 0.44** |
| 11 | –0.01 | –0.42 | 0.03 | – | –0.11 | –0.07 | 0.02 | 0.70** |
| 12 | –0.02 | –0.39 | 0.00 | –0.11 | – | –0.17 | 0.03 | 0.64** |
| 13 | 0.00 | 0.40 | –0.01 | –0.31 | –0.86 | – | –0.02 | –0.72** |
| 14 | 0.02 | 0.59 | –0.02 | –0.68 | –0.79 | 0.15 | – | –0.97** |

Residual effect : 0.06

**Significant at 1% level

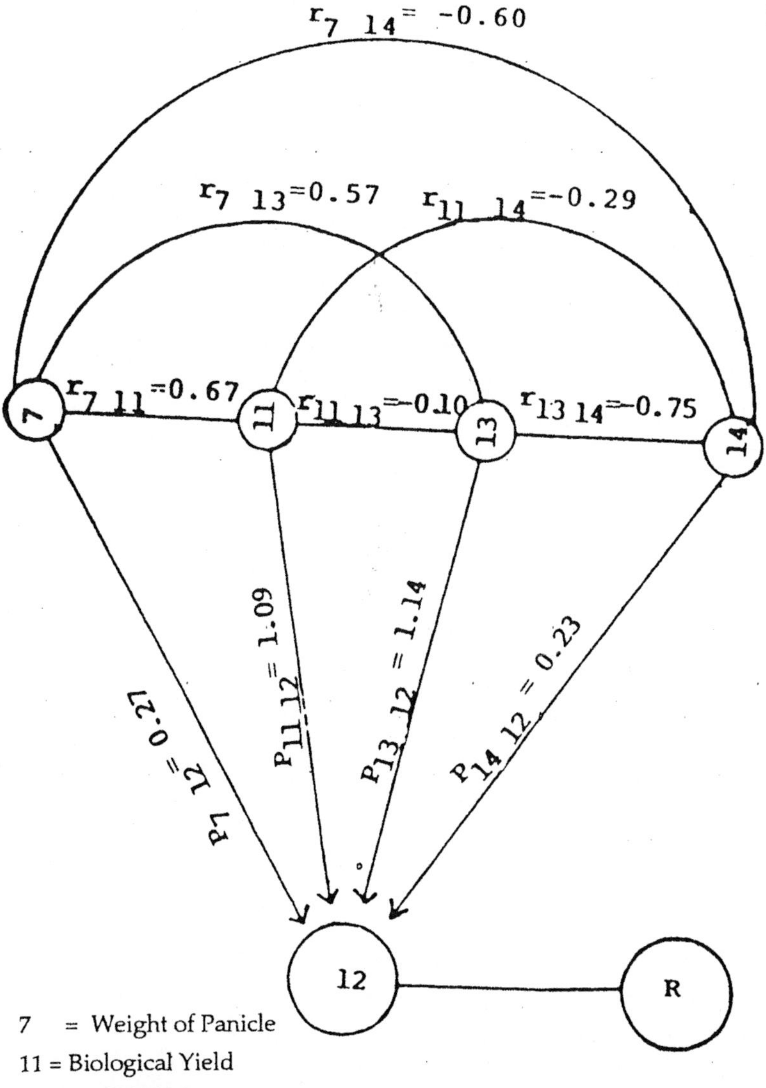


7 = Weight of Panicle

11 = Biological Yield

12 = Yield/Plant

13 = Harvest index

14 = Grain protein content

Fig. 1. Direct and indirect contribution of component characters in Setaria italica. A diagrammatic presentation.

(1977) for 1000-seed weight and panicle length; Randhawa *et al.* (1978) for ear length; Vishwanatha (1978) for panicle length and Nagarajan and Prasad (1980) for panicle length.

Harvest index showed significant positive correlation with yield as well as maximum direct effect. The indirect effects through yield/panicle, grain protein content and biological yield dropped the correlation value. Some what increase in corelation value to make it significant was achieved through positive indirect effects via weight of panicle, grain oil content and effective tillers/plant.

Strong positive correlation of biological yield with yield was mainly due to its high direct effect. This large direct effect was due to negative indirect contribution of this character via yield/panicle, harvest index and grain positive indirect effect through weight of panicle, 1000-seed weight and grain oil content.

Weight of panicle had positive and higher direct effect whereas its genotypic correlation coefficient was much higher due to its indirect effects through biological yield, harvest index, effective tillers/plant and grain oil content. Some of the positive indirect effect was compensated by negative indirect effect via yield/panicle, grain protein content, plant height and flag leaf area.

Grain protein content showed negative significant correlation with yield/plant but its direct effect was positive. This was mainly due to negative contribution of this character through harvest index, weight of panicle, biological yield and grain oil content. However, some of the negative indirect effects were neutralized by positive indirect effects via yield/panicle, plant height and days to maturity.

However, plant height, panicle length, peduncle length, flag leaf area, yield/panicle, 1000-seed weight and days to maturity exhibited positive correlation with yield due to indirect contribution of these traits via harvest index, biological yield and weight of panicle with relatively low or even negative direct effects.

Hence, the positive indirect effects for most of the characters were again largely through biological yield followed by harvest index and weight of panicle.

Low estimates of residual path ways 0.06 as indicated in the present investigation that the traits played an appreciable role.

Hence from the path analysis studies, it could be concluded that maximum direct effects were exhibited by harvest index, biological yield, weight of panicle and grain protien content. These traits showed strong genotypic correlation with yield and contributed to grain yield directly as well as indirectly. Therefore, due emphasis should be placed on these characters while breeding for high grain yield in foxtail millet.

## Selection Indices

Grain yield is a complex quantitative trait and is the multiplicative end product of many yield components. To make effective selection for higher yield, direct selection of yield components may not be as efficient as the simultaneous selection for several components. Therefore, relative efficiency of different selection indices over straight selection for grain yield per plant was analysed.

The characters included for construction of selection indices for grain yield/plant were as follows because these exhibited maximum direct and indirect effects on grain yield based on path analysis studies.

$X_1$ = Yield/plant

$X_2$ = Harvest index

$X_3$ = Biological yield/plant

$X_4$ = Weight of panicle.

The expected genetic advance of different selection indices were computed using five per cent selection intensity. For calculating relative efficiency of different selection indices over direct selection for grain yield/plant, the expected genetic advance for grain yield was considered as 100. Fifteen discriminant functions computed by using individual component characters and their various possible combinations along with expected genetic advance and their relative selection efficiencies over straight selection for yield are presented in Table 3.

Data indicated that the relative efficiency was low where selection was based on single component characters. However, maximum efficiency was exhibited for weight of panicle ($X_4$) 98.07% followed by biological yield ($X_3$) 69.00% and harvest index 57.37%. Mahadevappa and Ponnalya (1965) obtained highest relative efficiency for number of ear bearing tillers followed by weight of straw and number of fingers in ragi on the basis of single component characters.

Table 3 : Discriminant function, expected genetic advance of grain yield and relative efficiency of different selection indices in foxtail millet

| S. No. | Selection index | Discriminant function | GA | RE |
|---|---|---|---|---|
| 1. | $X_1$ | $0.71X_1$ | 1.47 | 100.00 |
| 2. | $X_2$ | $0.09X_2$ | 0.85 | 57.37 |
| 3. | $X_3$ | $0.13X_3$ | 1.02 | 69.00 |
| 4. | $X_4$ | $0.64X_4$ | 1.44 | 98.07 |
| 5. | $X_1X_2$ | $0.77X_1 - 0.02X_2$ | 1.48 | 100.41 |
| 6. | $X_1X_3$ | $0.66X_1 + 0.02X_3$ | 1.48 | 100.43 |
| 7. | $X_1X_4$ | $0.43X_1 + 0.31X_4$ | 1.52 | 103.47 |
| 8. | $X_2X_3$ | $0.11X_2 + 0.14X_3$ | 1.46 | 99.20 |
| 9. | $X_2X_4$ | $0.02X_2 + 0.59X_4$ | 1.46 | 98.87 |
| 10. | $X_3X_4$ | $0.03X_3 + 0.57X_4$ | 1.46 | 99.07 |
| 11. | $X_1X_2X_3$ | $0.68X_1 - 0.00X_2 + 0.02X_3$ | 1.48 | 100.43 |
| 12. | $X_1X_2X_4$ | $0.47X_1 - 0.01X_2 + 0.30X_4$ | 1.53 | 103.59 |
| 13. | $X_1X_3X_4$ | $0.41X_1 + 0.01X_3 + 0.30X_4$ | 1.53 | 103.59 |
| 14. | $X_2X_3X_4$ | $0.06X_2 + 0.09X_3 + 0.33X_4$ | 1.52 | 103.06 |
| 15. | $X_1X_2X_3X_4$ | $0.47X_1 - 0.01X_2 + 0.00X_3 + 0.30X_4$ | 1.53 | 103.60 |

Where

$X_1$ = Yield per plant
$X_2$ = Harvest index
$X_3$ = Biological yield
$X_4$ = Weight of panicle

Among combinations involving two component characters yield/plant and weight of panicle ($X_1X_4$) exhibited maximum relative efficiency followed by yield/plant and biological yield ($X_1X_3$) and yield/plant and harvest index ($X_1X_2$) as compared to harvest index and biological yield ($X_2X_3$), biological yield and weight of panicle ($X_3X_4$) and harvest index and weight of panicle

($X_2X_4$) and harvest index and weight of panicle ($X_2X_4$). Chinnaswamaiah (1975) indicated highest relative selection efficiency for the combination of length of main panicle and number of productive tillers in foxtail millet.

Among combination involving all three component characters (except yield/plant, harvest index and biological yield [$X_1X_2X_3$]) as well as combinations involving all the four characters exhibited high relative efficiency of almost equal magnitude. However, numerically highest relative efficiency was recorded over straight selection for yield/plant by an index of four component traits. On the basis of four characters combinations viz., 1000-seed weight, grain weight, length of ear and grain yield, Dhagot *et al.* (1977) obtained highest relative efficiency in *Setaria italica.* Hence, simultaneous selection for these four characters would help in increasing grain yield levels in foxtail millet.

## REFERENCES


Agolodia, A.V., Desai, K.B. and Tikka, S.B.S. (1979). Analysis of parameters of variability and interrelationship of yield attributes in finger millet. *Indian J. agric. Sci.*, 49 : 924-928.

Chinnaswamaiah, B. (1975). Formulation of selection indices through discriminant function technique for yield in *Setaria italica* (L.) Beauv *Mysore J. agric. Sci.*, 9:357-358.

Dewey, D.R. and Lu, K.H. (1959). A correlation and path coefficient analysis of components of crested wheat grass seed production. *Agron. J.*, 51 : 515-518.

Dhagot, N.K., Goswami, U. and Narsinghani, V.G. (1977). Character correlation and selection indices in Italian millet. *Indian J. agric. Sci.*, 47 : 599-603.

Dhanakodi, C.V. and Chandrasekhara, P. (1989). Path analysis in ragi (*Eleusine coracana* (L.) Gaertn.) for yield of fodder. *Madras Agric. J.*, 76 : 355-357.

Gasim, K.A.E. (1975). Correlation and path analysis in grain sorghum. *Cost African Agric. and Forestry J.*, 41 : 125- 133.

Gill, A.S. and Randhawa, A.S. (1975). Heritable variation and interrelationship in foxtail millet (*Setaria italica* (L.) Beauv.) *Madras Agric. J.*, 62 : 253-258.

Godawat, S.L. and Gupta, S.C. (1981). Yield components and their implication to selection in foxtail millet. *Madras Agric. J.*, 68 : 153-156.

Johnson, H.W., Robinson, R.F. and Comstock, R.E. (1955). Genotypic and phenotypic correlations in soyabean and their implication in selection. *Agron. J.*, 47 : 477-483.

Liang, G.H.L., Walter, T.L., Nickell, C.D. and Koh, Y.D. (1969). Heritability estimates and interrelationship among agronomic traits in grain sorghum. *Canad. J. Genet. Cytol.*, 11 : 199-208.

Mahadevappa, M. and Ponnalya, B.W.X. (1965). A discriminant function for selection for yield in *Eleusine coracana* Gaertn. *Madras Agric. J.*, 52 : 47-51.

Nagarajan and Prasad, M.N. (1980). Studies on variability, correlation and path analysis in foxtail millet (*Setaria italica*). *Madras Agric. J.*, 67 : 220-227.

Natarajan, U.S., Raveendran, T.S. and Appadurai, R. (1978). A path coefficient analysis of yield and yield components in proso millet (*Panicum miliaceum*). *Madras Agric. J.*, 65 : 430-434.

Patnaik, M.C. (1968). Variation and correlation in finger millet. *Indian J. Genet.*, 28 : 225-229.

Randhawa, A.S., Gill, A.S. and Asawa, B.M. (1978). Path coefficient analysis in foxtail millet. *Sci. and Cult.*, 44 : 134-135.

Reddy, C.D.R. and Lakshmi, K.J. (1991). Harvest index and yield parameters in foxtail millet (*Setaria italica* (L.) Beauv). *Indian J. Genet.*, 51 : 272-275.

Robinson, H.F., Comstock, R.E. and Harvey, H.P. (1951). Genotypic and phenotypic correlations in corn and their implications in selection. *Agron. J.*, 43 : 282-287.

Sandhu, T.S., Arora, B.S. and Singh, Y. (1974). Inter- relationship between yield components in foxtail millet. *Indian J. agric. Sci.*, 44 : 563-566.

Saxena, M.B.L., Rao, G.V.S. and Verma, R.C. (1979). Path analysis in *Panicum miliaceum*. *Indian J. Genet.*, 39 : 237-239.

Singh, K.D. and Rao, M.N. (1989). Association analysis in foxtail millet (*Setaria italica* (L.) Beauv). *J. Res. APAU*, 17 : 68-69.

Vishwanatha, J.K. (1978). Studies on genetic variability in world germplasm collection of foxtail millet (*Setaria italica* (L.) Beauv.). *Mysore J. agric. Sci.*, 12 : 519-520.

Wang, Z.C. and Han, M.S. (1981). Studies on heritability and correlation of major characters of *Setaria italica. Shanxi Agric. Sci.*, 10 : 9-10.

Wright, S. (1921). Correlations and causations. *J. agric. Res.*, 20 : 557-587.

Xu, Z.S. and Zhanes, Z.M. (1981). A preliminary study on the heritability and genetic correlation of the main characters in spring sown millets. *Shanxi Agric. Sci.*, 12 : 8-11.

Yadav, A. and Srivastava, D.P. (1976). Genotypic and phenotypic variability in *Panicum miliare* Lam. *Mysore J. agric. Sci.*, 10 : 185-189.